TV's

Not

Dead!

David Brennan

Published by New Generation Publishing in 2013

www.newgeneration-publishing.com

 New Generation **Publishing**

For my wife Jules, my son Shea and all who have had to put up with my mood swings during six months of arduous writing!

Contents

Introduction

PART ONE – THEN

Autobiography 3

History 12

Death 33

PART TWO – NOW

Performance 43

Why? 59

Engagement 65

Audience 70

Content 86

Context 94

Social 100

Trust 121

Payback 130

Online 145

PART THREE – SOON

Connected 163

Tellyporting 176

Entertainment 188

Advertising 204

Youth 220

Lessons 231

Final thoughts 251

References 255

Introduction

This book shouldn't really be necessary.

Ever since TV first began to elbow its way into the centre of our living rooms – and the centre of popular culture – it has attracted more than its fair share of criticism. It was too powerful, too influential, too disruptive or too shallow. But it was only after digital visionaries such as Nicholas Negroponte first opened our eyes to the digital delights in store for us that a new kind of criticism came to the fore. Television was too analogue, and therefore too vulnerable to the disruptive influence of digital technology. It couldn't compete with the personalised, interactive, ubiquitous nature of online technology. In short, television was dead!

Even before Negroponte's seminal work – *'Being Digital'*[1] – hit the bookshops, we saw a parade of technologists, futurologists, consultants and solutions providers dancing over television's grave. We would be freed from the tyranny of the network schedules, they confidently told us. We would be free to create and share our own content. We would never have to endure those tedious commercial breaks ever again. Television was finished!

With the possible exception of the financial sector, never have so many highly-paid, acknowledged experts got it so wrong. Here we are, well into the 21st century, over two decades on from the first predictions of TV's demise, and yet television has never been in better shape. People are watching more TV than ever before, they are paying more for it than ever before and they are being influenced by it more than ever before. The irony is, it is the very technologies that were meant to deliver the killer blows that are responsible for television's rude health. The lesson, if we are not to continue to repeat the same mistakes endlessly, is to not look at what technology can do but at what consumers want. It is a lesson that many of the digital fundamentalists have yet to learn.

In this book, I will be exploring the history of television, investigating its current health in forensic detail and then offering a perspective of its future within the emerging media ecosystem. This will be based on cutting-edge consumer research, including work I commissioned when I was at Thinkbox (the trade marketing body for commercial TV in the UK) as well as evidence from around the world.

The first section – 'Then' – provides a background to TV's current state, as well as explaining where the predictions of its death originated. I will set out my own experience of TV's history and some personal insights into why it continues to resonate so powerfully today. I will then provide a brief history of television in the UK. I should stress that I have chosen the UK not only because it is the market where I have worked for most of my career, but also because it is an excellent indicator for the future; it one of the world's most advanced digital markets, across both television and online. The UK provides a glimpse into the future; if TV can survive in the UK, it can survive anywhere.

The final part of this section will focus on the 'TV is dead' narrative; where it came from, the arguments put forward to support it and the accuracy of the predictions for the future of TV that had been thrown around like confetti.

In the second section – 'Now' – we look at where TV sits

in 2012, now that digital has become mainstream and the threats have become manifest. We will look at the state of the UK media market and television's place within it, as well as the component parts of the television viewing experience that have been totally underestimated or ignored by its critics. We will analyse the concept of engagement (and what that means in terms of audience, content and context) the power of the shared experience, the need for trust and how all of these combine to create value through payback – for the broadcasters, the audience and especially the advertiser. We will also look at the existing impact of online technologies on television's development.

The third and final section – *'Soon'* – will explore where TV is heading in the near future. A clear and meaningful picture is now developing of the mature digital media landscape, and TV's place in that landscape is being shaped, as households get to grips with digital convergence. I conclude this section with a few prescient lessons for broadcasters, marketers, platform operators and content providers. Oh, and the digital industry, which has such a poor track record of understanding TV's underlying resilience and opportunities. I will offer a few suggestions on how to get it right next time!

If they <u>had</u> been right, this book would have been an obituary – instead, it is a celebration of TV's revitalised role in the digital world we live in.

PART ONE

THEN

AUTOBIOGRAPHY

I was born in December 1955, just a few days after commercial television arrived in my part of North West England. We'd had the BBC of course, but it was ITV, the brash new kid on the block, that defined the true beginning of the modern television era in the UK. I was in blissful ignorance at the time, but the flicker of the box in the corner of the living room ('emotional central heating' as described by a 16 year old girl in a recent research study I commissioned) was a constant presence from the day my dad hauled a bulky second-hand black and white television set into our living room just 6 months after my birth.

Not that I ever complained. As a TV native, I could never imagine a world without it. Given a sherry or two at Christmas time, my aunts would show teary-eyed nostalgia for the times when all they needed for Yuletide fun was a piano and a pack of playing cards. For a teenager like myself, that sounded like a journey into hell and I'd turn back to watch *'The Morecambe & Wise Christmas Special'*, acutely aware of how fortunate we were to have access to such a magical source of entertainment. *In our own homes!*

I think we sometimes forget just how revolutionary a piece

of technology the television set offered. It provided 24/7 (well, more like 7/7 in the early days of limited transmission hours), instant access to an array of rich, audio-visual content; the American networks had already built a sizeable back catalogue of programmes to help fuel interest in the new medium, and the UK was quick to follow suit. Just like the birth of the internet, it provoked both instant attraction and general angst about its 'addictive' qualities in almost equal measure.

Vint Cerf, one of the founders of the internet, recently said during his Alternative McTaggart lecture in Edinburgh; *"imagine if the internet had been created first and then social media, and then television, think how excited we'd be"*. That sense of excitement in the medium itself has dimmed with familiarity, but the excitement its content generates can still be seen on those same social media every day.

Many of my childhood memories seem to involve TV. There are obviously the 'window on the world' memories – the moon landing, the deaths of Pope John XXIII and President Kennedy (absorbed with equal anguish in our Irish Catholic community), and the release of Nelson Mandela. There were the 'you had to be there' moments – England winning the World Cup, any one of Manchester United's numerous glory nights, seeing David Bowie appearing in drag on national TV for the first time. But there were also the 'pure TV' moments – praying *'Monty Python's Flying Circus'* would come up with something funny to stop my parents from constantly complaining about the lack of 'jokes', the dancing newsreaders on the 1976 *'Morecambe and Wise Christmas Special'*, the vicarious (and guilty) pleasures of reality TV and the heartrending tug of great dramas and documentaries.

TV quickly became enmeshed in our day-to-day routines. From my earliest memories, I can recall lying on the sofa watching *'Watch With Mother'*, praying it would be *'Tales of the Riverbank'* - which I loved - and hoping against hope that it wouldn't be those jerky, vaguely lobotomised puppets in *'The Woodentops'*, who inspired fear in me for years to come. Later, it was Friday night takeaways while watching *'Dallas'* together as a family, the ritual hiding-behind-the-sofa moments of *'Dr. Who'* and the early evening meals squeezed between the end of children's' programming and the start of *'Coronation Street'*.

During my teenage years, television took a back seat to the more alluring distractions of girls, bars, music and competitive sports. I did once run away from home (for seven whole hours!) because I wasn't allowed to watch *'Top of the Pops'* one night when I was fourteen years old, but otherwise I did what most teenagers have always done...and TV was never a big part of that adolescent lifestyle. I find it amusing that we still see teenagers' lower viewing of television signalling its impending decline; it has always been that way.

I never meant to go to University – in fact, it was only because my dad was dead set against the idea (he'd left school aged 14 and it had never done him any harm!) that I ultimately decided to go after a gap year mainly spent travelling the Middle East. I initially enrolled to study Economics and Financial Management in a futile attempt to make my university experience 'vocational', but I lost my vocation after just a couple of weeks of attending classical economics lectures that were often drier than unbuttered toast. I realised that this was not the subject for me; it was too dull, hypothetical and – to me at least – totally counter-intuitive. I never believed people behaved in the rational, linear, predictable ways the economists assumed. I switched to Psychology instead, which allowed me to indulge a lifelong fascination in people-watching and understanding human behaviour. I never for one moment expected that it would actually lead to a job, much less shape my career views for life, but it did.

At the time, Psychology, like the other social sciences, was desperate to be seen as a rigorous science, even while it was designated an arts degree. The Psychology faculty at my university – Sheffield – was dominated by the behaviourist theories of B. F. Skinner, which offered a reductionist view of human behaviour. Pretty much all of the psychological schools of thought for which I felt an affinity – Freud, Jung, the humanist Carl Rogers – were dismissed as being 'un-provable' and therefore 'unscientific'. It used to frustrate me, this sense that the complexities of the human mind could be as immutable and simplistic as testing and measuring changes in the physical world, but I had neither the means nor the skills to articulate those frustrations.

So, when I started thinking about my career options, about a week before my finals, using Psychology as a key part of that career was quite a way from my mind. I knew I wanted to work in 'telly' in some capacity, but had no idea what capacity that would be. In the end, my careers adviser suggested I was being naïve by thinking I could get into television and perhaps I should consider a career in accountancy or insurance instead. Fortunately, I ignored her advice and set about inundating the television networks and production companies with increasingly desperate job applications. Two years later, I got my first job in television with Ulster TV as an Audience Research Executive, analysing ratings, advertising spend and consumer behaviour.

It was something of a surprise when I first started working in TV, and found out that the people running the industry, and those making the big money decisions about it, weren't as enthusiastic about the medium as I was. I'd scan the weekly 'My Media' column in Campaign magazine and marvel that industry leaders could find time to read several serious newspapers every day but would rarely get time to watch TV, apart from news and the occasional documentary. I'd present to senior managers commanding mega marketing budgets who would have no idea of the programmes that were attracting their consumers in their millions! I once earnestly briefed a head of programming at HTV about what I could tell him about the audience to his programmes, only to be told *"I make my programmes for seven people, and I doubt any of them will be on your research panel"*.

This has been an infuriating aspect of my career in media. Television has always had a central part in most people's lives, and yet those who need to understand that audience have often seen them as something different, inhabiting an alien world where *'Coronation Street'* is more important than The Times leader column and conversation is more likely to be about *'The X-Factor'* than the economy.

I worked within ITV for the next 15 years, running research and marketing departments for several regional franchisees, followed by a couple of years as Controller of Audience Planning for the ITV Network Centre, during its most successful ever ratings period (1993-95). Television then was 'comfortable'; not necessarily the *"licence to print money"* as described by Lord Thompson during the late 1950's but certainly profitable, growing steadily and virtually unassailable in its role at the centre of home entertainment.

Comfort often brings complacency. That was brought home to me at an ITV network strategy workshop I attended in 1994. I had been briefed to present my view of the shape of the television market in ten years' time. I duly went through my list of opportunities and threats, made a few projections and arrived at the conclusion that a worst-case scenario could see ITV's share of viewing decline to 23% by 2004. At the time, ITV was riding high with an all-time share of 39% and the two looming threats I saw coming – a resurgent multi-channel offering and a relative lack of sustainable content – were at the periphery of its radar. My projections were summarily dismissed. Then again, I was out in my prediction by two years – ITV hit the 23% mark in 2002, two years earlier than I had estimated!

In 1995 I received a call out of the blue – would I be interested in a new VP Research role for up and coming pay TV player Flextech Television? Logically, it would have meant a step down from the biggest commercial broadcaster in the UK market to one of the smallest (Flextech's combined channel share, including Discovery channels, Bravo and The Children's Channel, was less than 1%). It meant moving from the secure, established comfort of network television to a position on the outside, working with a challenger brand. I took the job.

This new world had a different set of rules, which appealed to me. During my first management meeting, we were discussing the schedules of one of our channels that had struggled to attract an audience. I mentioned that most pay TV audiences simply didn't know what was on the minor channels and the US approach of stripping and stranding schedules, where the same programme appears in the same slot each day and certain schedule strands can be themed to link programmes together, seemed to work better for specialist channels. If I'd still been with a network broadcaster, that idea would have taken several months to reach the advanced discussion stage; in this case Joyce Taylor, our MD at the time, paused for maybe five seconds, and then uttered my three favourite words. *"Let's do it!"*

The late 1990s were a giddy time for media technology and television was right in the centre of it. We were looking at all of the impending technological developments – wireless broadband, 3G and 4G mobile technology, smartphones and PDAs – and working out how our TV businesses could use them to generate, enhance, distribute and integrate our content. The idea that the internet would somehow replace television was still a dotcom crash away from becoming orthodoxy.

In 1999 I was promoted to also take charge of channel development, unfortunately just at a time when traditional sources of funding – especially subscription revenues – were making most new channel propositions completely unviable. Still, the growth of the internet and the interactive possibilities of digital TV meant there were plenty of avenues of development to explore. During that time, I was involved in the launch or re-launch of dozens of channels, but it was the new ways in which television could be delivered that really excited me. One example – Living Health – gave me a deep insight into how the role of television, and the unique place it has in most people's media usage, could deliver more than just entertainment.

I led the team behind Living Health, a government-funded live pilot to show how healthcare could be delivered via digital interactive TV. It ran for 6 months in Telewest's Birmingham Cable area and achieved some of the best per capita user stats for a purely interactive TV service ever recorded anywhere in the world. The service provided information, advice, access to support services and even a video link to an NHS Direct nurse if needed. During its six month run, Living Health was tried by over 85% of Birmingham Cable's subscriber base and achieved record penetration, satisfaction and loyalty ratings. Despite winning a European Multi-Media Award (EMMA) in 2001, Living Health never attained permanent government funding and had to be mothballed, but it gave me a very early taste for how television could act as a fantastic template for similar cutting-edge services because it was so easy to use, familiar, trusted and focussed on delivering great audio-visual content.

Then came the dotcom crash, and television stumbled, along with thousands of internet start-ups and most other media businesses. Investment in new channels and interactive services dried up and television became...well, just television again. Much of the attention was on the continued march forward of pay TV and the emerging threats from digital television recorders (DTRs– often known as PVRs), such as TiVo and Replay TV.

Meanwhile, the dotcom industry sought salvation in Web 2.0 while renewing its predictions of TV's demise ever more vehemently. I will cover this 'TV is dead' narrative later in this section. Instead of exploring what digital television could deliver, attention focussed on what television couldn't deliver, compared to the personalised, interactive, portable and viral properties of the web. Most of television's established benefits – efficient distribution, shared viewing, trust and familiarity, compelling content and strong channel brands – were dismissed as anachronistic or irrelevant.

In 2006, I joined the newly-formed Thinkbox, the UK commercial television industry's belated attempt to market the medium, as Research & Strategy Director. We immediately put in place a programme of research, all of which informs this book and much of which is directly quoted. I soon realised that TV needed something inspiring and thought-provoking; the usual research cocktail of survey research or focus groups would simply not be enough, given the new insights into how the human brain operates, the power of emotional and implicit, sub-conscious processing and the emergence of behavioural economics, that were turning established ideas about how advertising works on its head. We attempted to tell television's 'story' through a new lens.

Our use of innovative research techniques allowed us to see beyond the false claims and self-representations of traditional market research. We employed ethnography, biometrics, implicit attitude testing, neuroscience, macro and micro econometrics, conjoint analysis and creative workshops to really understand what was going on beneath the surface. I took a lot of risks commissioning such leading-edge research methodologies because I usually had no idea whether they would yield positive or negative results for TV. Needless to say, television never let me down. It always seemed to come out on top, whichever measures we used. More importantly, the insights that these new techniques provided helped to explain a lot that had been previously invisible to us when we had relied on the old measurement techniques. This book is my attempt to bring it all together to explain why, despite the widespread conviction that we were witnessing its death throes, TV has defied all of the obituaries to come back fighting.

I have spent more than three decades happily providing research insights across all areas of the TV business – including advertising sales, marketing (trade and consumer), programming, scheduling, new product development and strategy. During that time, I have seen many peaks and troughs ripple across the industry, caused by economic conditions, poor industrial relations, regulation, competitive activities and, occasionally, poor management.

There have been times in the last decade or so when TV's future has been less than secure. But, not only has TV survived, it has survived in a very Nietzschian way; what couldn't kill it has certainly made it stronger and even those who so confidently predicted its death are finally recognising this to be the case.

How could they have got it so wrong, so spectacularly and so consistently? What did they miss, in their five year predictions of the complete replacement of broadcast by IP, channels by aggregators, schedules by on demand and commercial viewing by fast forwarding? As this book will show, they got it wrong because they looked at technology, when really they should have been looking at people. The reason for TV's continued strength isn't about shiny new gadgets or technological innovation so much as established behaviour patterns and emotional attachments; things which affect behaviour far more fundamentally than any technology can. If we don't learn these lessons now, we can only get it even more spectacularly wrong in the future.

TV is back and, 30 years after my career began, things finally feel like they have started to come full circle.

HISTORY
INTRODUCTION TO THE THREE AGES OF TELEVISION

I initially developed the idea of the three ages of television when I worked for Flextech Television, the European content arm of Liberty Media, in 2001. I'll admit that part of my reason for doing this was mischievous; the online community had just begun to talk about the revolution of Web 2.0 and I wanted to stress that television was going through its own revolution, even at a time when most digital commentators still saw it as a traditional, limited medium. In a childish way, it was a way of saying *"we have more ages than you do, na-na, na-na-nah!"*, but there was a serious point behind it; that television was itself going through massive innovation, primarily as a result of its position as the dominant digital entertainment technology. Just about every home in the UK is now a digital television home, and that has helped to unleash a torrent of new products and services that have all had an impact on the way we access, watch, participate in and respond to the television experience, although maybe not nearly so much as has been predicted.

The three ages of television apply to every TV market I have investigated, although each has its own unique pace of change and industry landscape. Although I will frequently refer to other markets, most of the evidence and insights will come from the UK market. The reason for this is twofold; many of the breakthroughs in our understanding of TV's emerging role have come from the UK (especially Thinkbox), but also the UK is the most advanced digital economy in the world. Consider these facts;

1. It has the highest proportion of online shoppers and (per capita) the most spend via online retail in the world.

2. It is also the most export-positive digital economy in the world, exporting almost three times what it imports via online transactions

3. The UK population is the most active in social media, with more Facebook members per capita than anywhere else in the world.

4. The UK is Google's second largest market after the USA and per capita generates more revenues and search activity than anywhere else.

5. It has the second largest online advertising market in the world after the US and, in share terms, online takes the highest share of total advertising revenues globally

6. It is the world's biggest internet gateway, with more than a third of all global internet traffic routed through UK servers[2]

7. It is set to be one of the first countries to switch over completely to digital television transmissions – it will become totally digital in 2012.

8. It has been a world leader in TV exports (almost half of all worldwide TV formats emanate from the UK) and

technological innovation (e.g. Sky+, red and green button interactivity, on demand television and 2-screen content applications)[3]

All of these trends, however, have only become apparent in the last decade or so, aided by UK attributes such as nature of the economy, density of population, language and existing media landscape.

It all seems a long time ago since television first staggered out of the laboratory and into our living rooms, but as it has evolved through its three ages, TV has undergone changes that have both transformed and reinforced its position as our primary entertainment medium.

It hasn't always been that way; ever since TV became established, there has been a chorus of voices predicting its demise. It has often sounded like a mission to rid the world of a corrosive influence. For others it has been summary rejection of TV's ability to survive in the face of such a technological onslaught. For many more, it has been amazement that people would want to sit back and be passively entertained when there are so many more 'useful' activities they could perform instead.

It's been a long process to get from there to here, but it helps us to understand just why TV is so ingrained into our daily rituals and conversations and why it cannot simply be replaced by supposedly better technologies.

TV is here to stay; its evolution has ensured its survival.

THE FIRST AGE OF TELEVISION

The first age of television flickered into being from the moment the laws of electro-magnetism were first discovered in the early 19th century, and really began to take off in the aftermath of World War 2. Although that encompasses a long time period, I see that age primarily as the prologue to the first age of television, which really began when a coherent commercial market began to develop in the early 1950s. In the UK, it can be traced back to a specific date – September 1955 – when commercial television was launched, and in many other developed markets the start date was within a few years of the UK, underlining the global and immediate nature of this fledgling technology.

The first age of television can be defined by a number of characteristics. These include;

- **Limited channel availability** – owing to both limitations of the available spectrum and the high cost base required to set up a fully-functioning broadcast operation in those early days of TV. Most national markets were able to sustain just four or less channels – a mix of regional and national, state-owned and advertiser-funded

- **Limited set functionality** – for most of the first age of television, TV sets could receive TV channels and...well, that was about it. During the latter part of this period set reception (from black and white to colour), interaction (remote control technology) and enhanced broadcast (teletext services) evolved, but for most viewers most of the time during this period, the best they could hope for would be switching between broadcast channels to find the content they wanted to watch

- **Linear programming** – the phrase 'appointment-to-view' television was coined during this period, because the viewer had to make the appointment to view at the time the programme was scheduled. Although the VCR became commercially available during the late 1970s and achieved mass market penetration towards the end of the first age of television, these were often used for pre-recorded content. Timeshifted viewing of linear schedule content never exceeded 5% of all programme viewing, even by the late 1980s.

- **Highly regulated markets** – it is strange that, ever since it became widely available, television has been perceived as a trivial, dumbed-down, *inferior* technology, and yet from day one it has been more highly regulated than any other media platform before or since. Regulators initially obtained their power via the awarding of broadcast licences; the commercial potential in television meant demand for licences generally exceeded supply, so regulators could subject broadcasters to strict rules, involving advertising, scheduling, programme content and ancillary activities.

Television's Pre-history

The history of television can be traced back to a modest, wood-panelled lecture theatre at the University of Copenhagen in 1820. Christian Oersted, was preparing to give a lecture, when he chanced upon a discovery with ultimately startling consequences. As he passed an electric current through a voltaic pile (an early form of battery) he noticed that the needle on a nearby compass would flicker and change direction, returning to its original position only after the current had been turned off. Oersted became convinced that he finally had the proof of the relationship between electricity and magnetism, and immediately published his findings. The publication of Oersted's findings set off a wave of investigation into this significant phenomenon, leading to the electricity revolution, the transport revolution, the communications revolution...and The

Simpsons!

The first recorded instance of the use of the word 'television' to describe this new technology was at the 1900 World Fair in Paris. Initial demonstrations were presented by John Logie Baird in 1924 and the first transmission of pictures (of US President Herbert Hoover) across long distances was conducted in 1927 by the Bell Telephone Company, leading Hoover to state;

> *"Today we have, in a sense, the transmission of sight for the first time in the world's history. Human genius has now destroyed the impediment of distance."*

A year later, the world's first US television broadcasting licence was awarded to Charles Jenkins and in 1930 his station had transmitted the first ever TV commercial and created the earliest TV character (Felix the Cat). Meanwhile, in the UK, the BBC had begun TV broadcasts, initially to just a few hundred households around the London area.

Opinion was still very much divided regarding television's prospects. Although an increasing number of households began to buy the new TV sets, its potential was still seen as a communications or monitoring device, and initial interest was much higher among the telecommunications companies and the military than from within the entertainment industry. Indeed, other parts of the media industry were noticeably cool about this new contender. Renowned journalist C.P.Snow, speaking in 1932, was dismissive of TV's potential;

> *"Television? The word is half Greek and half Latin. No good will come of this device"*

Meanwhile, the New York Times leader from 1939 confidently stated;

"TV will never be a serious competitor for radio because people must sit and keep their eyes glued on a screen; the average American family hasn't time for it."

Newspapers and radio broadcasters attempted to stifle the new medium – for example, by not allowing them access to their news-gathering resources – while publicly denigrating its chances of survival, but just as TV appeared to be building a foothold, especially in Europe, it was interrupted by something much more important; the Second World War.

TV Goes Commercial

It is amazing that, amidst the destruction of post-war Europe, television immediately became part of the rebuilding process. By the mid 1950s, TV was established across the continent, as well as North America, Australia, Japan and even communist USSR. Following the War, the success of the US advertiser-funded model together with a pent-up demand for more entertainment to lift the post-war gloom, led politicians to consider how to respond to the growing popularity of the new medium.

Early TV networks were either publicly funded or relied on advertising, sometimes a hybrid of the two. Critics were either indifferent or contemptuous and the audience did not appear much more enthusiastic at first. Very quickly, however, TV reached mass market penetration and revenues began to move upwards.

By the mid 1960s television across Europe and America seemed to be in strong health, both financially and in terms of viewer popularity. A growing audience base and an increasing portfolio of advertisers meant revenues across the industry were reaching substantial levels. Both were helped by the increasing disposable incomes of the post-war generation.

The two key trends during the 1960s and 1970s were greater

regulation and improving technology. Regulators were aided by increasing demand for TV franchise licences and a scarcity of spectrum, enabling them to dictate programme and advertising content. In the UK, cigarettes became the first banned TV advertising product in 1963.

TV technology was boosted by the first long-distance satellite transmissions in 1962, the introduction of the higher definition UHF signal in 1964, and the introduction of colour television in 1969. The mass adoption of remote controls from the late 1970s has also been credited with changing the nature of television viewing.

But it was how technology helped the broadcasters to make content more engaging that is perhaps the real story of TV's early years. Through a combination of more portable and higher quality cameras, more functional studios and improving post production facilities, the early 'static' programmes from the networks were beginning to be replaced by live transmissions, outside locations and more seamless, immersive storytelling. Overall, audiences increased consistently through the 1960s, averaging around a fifth higher than they had been at the start of the decade.

The money followed the viewers, at least until the end of the 1970s. This helped create the conditions for investment in the technology and content, but there were some negative consequences. For example, industrial action was a key feature of television in the 1970s, and TV broadcasters were often accused of complacency, operating as they did with limited competition. In the UK, ownership of a TV franchise was described by Lord Thompson of Fleet as *"a licence to print money!"*

The launch of new channels during the 1980s helped to make television more relevant for some severely under-served audiences, especially the young. The launch of Channel 4 in the UK enabled landmark series such as *'The Tube', 'The Word'* and *'Comic Strip Presents'* to inhabit peak time slots. Around the same time, RTL

Plus and Sat 1 launched in Germany (1984), M6 and La Cinq (1987) in France and Canale 5 (1980), Rete 4 and Italia 1 (both 1982) in Italy, often doubling the number of channels available at a stroke. New terrestrial licences were being awarded right up to the end of the 1990s (Channel 5 in the UK was launched in 1997), but by then the introduction of another national free-to-air channel was an anachronism. TV was well into its age of transition, its first major disruption since the launch of commercial television almost four decades before. The second age of television was dawning.

THE SECOND AGE OF TELEVISION

The second age of television can be characterised as the explosion of channel choice, as new platforms, especially cable and direct-to-home (DTH) satellite, allowed broadcasters to bypass the limited terrestrial airwaves. In some countries – for example the USA – cable started early, primarily as a means to provide network transmissions to communities which could not receive over the air. In these markets, the distribution infrastructure was, by the 1970s, sufficiently established to make it viable to provide specific channels and services on top of those provided by the networks. In other markets, including the UK, where terrestrial reception was almost universal, multi-channel penetration was much lower and it took a new distribution network (DTH satellite) and often a significant entrant into the market (Sky, Mediaset, Canal Plus) to really kick start the pay TV revolution.

In the UK, the second age of television can almost be defined by an exact date – the merger of Sky Television and British Satellite Broadcasting – as that is when many of the market conditions mentioned above came together in a unified way. Its end date, however, is harder to map out precisely.

The main characteristics of the second age of television are;

- **Viewers paying for something that had always been free.** Although it is only 20 years or so since pay TV really began to take off, at the time the idea that people would pay for television was commonly felt to be fanciful at best.

- **A pay TV oligopoly.** One or two new entrants to the market, providing both distribution infrastructure (usually at high capital cost) and the premium programming to create the conditions for sustainable household penetration levels.

- **An explosion of channel choice,** -as new channels, both global and local, take advantage of the newly available spectrum.

- **The beginnings of on demand** – although it could be argued that the VCR became widely available during the first age of television, it reached mass penetration in line with the growth in pay TV penetration. Meanwhile the pay TV operators themselves began to experiment with NVOD (near video on demand), using available channel spectrum to provide an 'almost' on demand service (usually the same programme shown at staggered start times).

- **Fragmentation of audiences** and a steady migration away from the traditional terrestrial networks.

It was only when cable-exclusive channels began to appear in the USA, following deregulation in 1972 that a true added value pay TV market could develop. The first basic cable networks included HBO and Ted Turner's WTCG in the mid 1970s. Premium networks based around movies and sports soon followed. This set the scene for pay TV expansion into other developed TV markets. Initial offerings were very limited, especially outside the USA. The early Sky Television service, for example, consisted of just four channels and even the merger of Sky and BSB in 1990, credited with boosting the pay TV market in Europe, started with a mere five channel service. But the biggest deal was yet to come. British broadcasters had only recently begun to screen live football matches on a regular basis, but when the rights were up for negotiation in 1991, BSkyB won through a joint bid with the BBC (which was anxious to retain rights to the highlights and retain a foothold with live football). The bid was £304 million, considered an excessive amount at the time, but it is now recognised as the deal that saved the company and kick-started the UK's pay TV market.

Not only did the Premier League deal produce an immediate and sustainable increase in subscribers, it also enabled BSkyB to generate a business model based on an encrypted, premium channels offering. The business model was replicated in France (Canal Plus), Italy (Sky Italia) and across Europe, leading to rapid growth in pay television subscribers throughout the 1990s.

The growing scale of pay TV led to an increase in the number of channels offered – often 10-20 times what terrestrial viewers had been used to – which was boosted further with the launch of digital pay TV services in the late 1990s. These were initially focussed on providing even more channels, through digital's greater channel capacity, but innovations such as the Electronic Programme Guide (EPG), interactive services (such as BSkyB's Open...) and enhanced broadcast features (e.g. Sky Sports Extra) began to also offer viewers more control and convenience.

All of these factors had a part to play in digital TV's rapid adoption, especially in the UK. In the three years from 1998 – 2001, household penetration (i.e. the percentage of UK homes having at least one digitally-enabled TV set) grew to 40%; to put that into perspective, to reach the same penetration level, it took the VCR **five** years, colour TV **eight** years, the PC **ten** years; even broadband internet penetration took **six** years (from 1999 – 2005) to achieve that same penetration level.

The pay TV market showed an immediate spurt of growth following the beginnings of digital TV. In mid 1998, a little over 6 million households subscribed to pay TV services in the UK, but by the end of 2001 that had hit the 10 million mark, more than 80% of those subscribers opting for digital. This was one of the main reasons why the UK government's target for converting the whole of the UK from analogue to digital broadcasting by 2012, considered wildly optimistic when it was first announced in 1997, has in fact been comfortably reached.

The headline impact of all of this activity was the effect on the main channels' share and especially the performance of the top-rating programmes, which could no longer guarantee the audiences they had traditionally delivered. That was when the negative headlines about audience fragmentation began to bite and the 'TV is dead' narrative really began to take hold.

Meanwhile, we experienced around fifteen years of the second age of television. It may now seem to have been a transitory stage, compared to the sedate passage of the first age and the tumultuous beginnings of the third age. The second age of television transformed the economics of the industry, demonstrated how much people were prepared to pay for the right TV content and opened the way for television to evolve into much more than a limited choice, passive entertainment device.

The irony is that, when many in the digital industry were expecting digital technology to kill television, in fact it transformed the medium, and without the second age of television the industry may never have been equipped to cope. Meanwhile, we were entering the third age of television

THE THIRD AGE OF TELEVISION

The obvious thing to do would be to set the start of television's third age as the launch of digital television broadcasting, but in my mind that would be too soon. That is because the TV companies were initially more focussed on using the increased capacity (digital TV could fit up to six channels into the spectrum required for a single analogue channel) to offer more channels. That made the early days of digital television feel more like an acceleration of the multi-channel explosion that characterised the second age of television.

It is true that the early days of digital provided a step change in interactivity, introducing the Electronic Programme Guide (EPG), as well as the first red button applications. It could be argued, though, that these were functionally similar to previous innovations such as the remote control and teletext.

The changes which have since ushered in the third age of television comprise so many new innovations and applications that it feels like we are going through a new phase of innovation every month. Television is evolving so rapidly it's hard to keep up but, as we shall see in later chapters, the main thrust behind the advent of TV's third age is the technology that was meant to destroy it; the internet.

The growth in online penetration, usage and functionality has transformed television, not just through the new ways in which television programming can be distributed and accessed, but also by how broadcasters, advertisers and the viewers themselves can add value to the viewing experience. I am therefore going to date the start of the third age of television as the date when broadband penetration hit the 50% level.

More than half of the UK's households were broadband connected by November 2006 (several months before the US reached the same level). With an established digital television infrastructure in place and broadband (especially wireless broadband) freeing up the laptop and other switched-on devices for more casual forms of use, this has been the tipping point which can now be seen as the beginning of the third age of television. The main characteristics of this third age are;

- **Ubiquitous access to content** – the concept of 'Martini TV' (anytime, anyplace, anywhere) is a key feature of TV's third age. Once a programme has been broadcast (and in some cases, before broadcast has even occurred), viewers can access that content on a range of devices. This means they can watch at a time to suit them (time-shifting) and/or in a place other than the room in which a TV set is located (place-shifting, or mobile/portable TV).

- **Enhanced broadcast features** – lots of additional content contained within the broadcast stream (e.g. programme information, advertiser information), accessible either as side information or via the red button.

- **True interactivity**, allowing a truly personalised viewing experience and a full range of response features for both editorial and advertising content.

- **Convergence of screens and devices** – so that companion devices (such as smartphones, laptops or tablets) can independently co-ordinate with what is on the main TV screen and provide ancillary content (audio-visual or otherwise) or response opportunities.

- **A greater focus on convenience and 'packaging' of services** to optimise the TV viewing experience - especially from the pay TV operators, who shift from more choice to greater convenience in terms of their strategic focus.

Television 3.0 – what doesn't kill us makes us stronger

Until the dotcom crash of 2000, the consensus was that television would benefit greatly from digital technology in terms of added value for both audiences and advertisers, as well as offering up a potential gold mine of transactional revenues.

Broadcasters have always been accused of coming to the digital party late and unprepared, but in the UK the BBC and the pay TV players had been active since the late 1990s. The BBC began to make waves with the launch of the BBC website in 1997, which proved immediately successful, to the dismay and consternation of the newspaper industry. The BBC also pioneered interactive television via the red button in 1999, mainly as a replacement for Ceefax, the BBC's teletext service. It has since gone on to be one of the most successful red button services worldwide and, for many BBC viewers, works as an alternative to the internet (up to a third of BBC viewers are not regular internet users or don't have access at home). This level of innovation and quality standard-setting possibly reached its peak with the launch of the BBC iplayer in 2007, which transformed the on demand TV market in the UK.

The pay TV operators were also very active at the turn of the millennium. Since 1998 they had seen pay TV penetration levels soar. It was time to move the focus away from subscription drives and more towards expanding the average revenue per user (ARPU). The smart money was on transactional revenues.

It is difficult to believe now, but red button revenues in 2000 were seen as the next stage in TV's evolution. Even the analysts (many of whom started to predict the death of television just a few years later) were often bullish about the power of TV combining with the opportunities for direct response; Forrester enthused about the huge potential of 'lazy interactivity' and many other analysts started to talk up the transactional television marketplace. The concept that people could see something of interest – via an advertisement or programme – and then immediately respond for more information, a product sample or even a direct purchase, was perceived as a potential game changer for TV.

In retrospect, as with many aspects of the digital revolution, the hype was not matched by the consumer experience. Red button interactivity was slow, clunky, expensive and, for advertisers at least, less than compelling. It was successful at opening up new programme content (e.g. extended coverage of major sporting events) but, unless advertising content was extremely entertaining, viewers appeared unwilling to switch out of the programme they would be watching in order to interact or respond to an advertiser's communication. Very rapidly, the potential for transactional TV was significantly marked down and the gaze of the technologists fell elsewhere.

The revolution will not be televised...it will be blogged about instead

While all of this was happening, online was taking off. A global user base of around 18 million people in 1995 had grown to 360 million by the end of 2000. This figure, coupled with faith in the power of broadband to fuel future growth, created a frenzy of worldwide speculation. Massive investments on what are now considered unsustainable financial projections became the norm. Anything with dot-com in the title appeared to be able to attract funding with ease.

All of that changed on March 13th, 2000, when a raft of IPO stock options entered the market simultaneously, triggering a chain reaction of selling. This led to a fall of over 3% in the NASDAQ value in just one day and by October 2002 the NASDAQ was worth less than a quarter of its peak value. It was like many of the previous stock market bubbles – from the South Sea Island bubble of the 18th century right the way through to the US housing market bubble of just a few years ago. As Warren Buffet, the famous US investor and philanthropist noted;

> *"after a heady experience of that kind, normally sensible people drift into behaviour like that of Cinderella at the ball. They know that staying on at the festivities will normally end in pumpkins and mice"*

The now familiar aftermath was a large number of dotcom bankruptcies. Less well-reported was the impact on many tech-based companies with sustainable and established business models. It was not just Time Warner, panicked into the merger with AOL that has been called 'the worst business deal in history'[4], that suffered badly. Most TV companies suffered the chill wind of investor caution coupled with advertiser sentiment impacting on their short-term and longer-term profitability.

In the UK, between 1980 and 2000, television advertising revenues had more than doubled to £3.5bn, more than twice the rate of GDP growth over that time. Suddenly, in line with the dotcom crash, revenues got stuck –initially stalling and then, from 2005 onwards, going into freefall, recording a 13% decline in just four years.

From cyclical to systemic – TV's long-term hangover

TV revenues had flat-lined before, but the declines in advertising revenues from 2005 were the worst that TV had ever experienced, not just in the UK but in most other developed TV markets as well. As if this wasn't bad enough, in the middle of the decade the additional burdens placed by the UK regulators as a condition of the creation of a single, unified ITV system added fuel to the fire.

Contract Rights Renewal (CRR) effectively required that ITV's prices could only rise proportionately to the network's audience share. Given that ITV had failed to increase its audience for more than a decade, this appeared to be a huge gamble. As so it proved. Unfortunately, the impact was to also reduce prices for television overall, as the market was largely benchmarked on ITV's price.

Adding in the impact of economic events, the downward pressure on television advertising revenues became a major problem. Price fluctuations might have been perceived as cyclical, if it weren't for the fact that the medium was also seen to be facing a newly-emergent threat from the online sector.

The online industry had shaken off the hangover from the dotcom crash and began to attract major investment once more. In particular, the highly personalised, social and interactive experiences that broadband allowed meant that 'Web 2.0' was quickly becoming a reality. Meanwhile, broadband's greater capacity meant that rich media could be delivered, providing more TV and TV-like content online. Broadband penetration began to climb rapidly, to the extent that almost three quarters of the UK population were connected by the end of the decade[5].

All of this helped to fuel the crisis in TV revenues, sparking a number of press reports and industry sentiment based around the idea that 'TV is dead'. The online industry was quick to claim both credit and cash. During the same period online advertising revenues began to increase rapidly. From less than £20million revenues as recently as 2002, the internet started to overtake radio and outdoor (both in 2004) and pushed through the £1billion level the following year. More than four fifths of that increase could be directly attributed to one media brand - Google.

The rise of Google has already been documented in numerous books and articles, but the improved web search experience, coupled with their advertising auction model, has had a significant impact.

In 2000, Google began to sell simple text-based advertising around key search words or phrases on the basis of click-through rates and a bidding mechanism. The ingrained efficiency of this system created a revenue bonanza, reinforcing Google's dominance in the internet search market – by 2004 the company was responsible for almost 85% of all global searches.

In the UK, Google is now officially the biggest advertising media channel. For many of the digerati, Google has been the catalyst, paving the way for the internet to become the dominant advertising medium. In its wake would come revenues for a wide range of online advertising platforms.

It certainly appeared that way in September 2009, when the Internet Advertising Bureau (IAB) announced that the UK had become the first global market to see internet advertising revenues eclipse those of television. In the six months to June 2009, internet revenues exceeded £1.75 billion, a 5% annual increase (at a time when ALL other media were seeing major decreases in spend – TV saw a 10% fall that year), almost 20 times the revenues from eight years earlier. As The Guardian newspaper commented at the time;

> *"The milestone marks a watershed for the embattled TV industry, the leading ad medium in the UK for almost half a century. It has taken the internet little more than a decade to become the biggest advertising sector in the UK."*

It certainly was a watershed. The dominance of search

advertising in online's figures created the biggest shift between display revenues and classified revenues since advertising first began. (Display advertising is best described as advertising that uses a range of techniques to display the brand or product to the consumer, and is usually about brand building. It is distinguished from classified advertising, traditionally a print phenomenon, which usually appears in defined sections and offers products or services direct to the consumer).

There has long been a debate about exactly what category of advertising Google can be placed within, but most would agree it is primarily a form of classified advertising (although some have suggested it is more akin to point-of-sale). Either way, the shift in revenues away from brand-led display advertising was significant.

The extent of display's problems can be seen in the failure of online display to break through. Even during the year when online claimed to overtake TV, online display revenues were down 5% and there was a real sense that the display advertising model – the foundation of commercial television in the UK – was bust. Media agencies were looking at new models, mainly motivated by the opportunity to by-pass paid-for display advertising as much as possible in favour of 'free' advertising created by viral content and the increasing power of word of mouth via social media, as well as more response-led advertising. So, when TV advertising revenues tumbled by £500m in just four years, it was predicted to be the end of the line. TV was on its last legs. The digital industry vaunted the new possibilities of content freed from schedules, allowing participation rather than the mere consumption of entertainment and the much more relevant and personalised content that online could deliver. In short, TV was, by its nature, inherently doomed.

The TV broadcasters had reacted to the unthinkable by establishing Thinkbox, commercial television's trade marketing body, in the summer of 2006. It was an irony not lost on the trade press at the time that television companies had taken more than 50 years to begin to actively sell the medium, whereas newspapers had been represented for all of those 50 years, radio for 25...and even the internet had a 10 year head start on TV (the UK's IAB launched in 1996). But there was already a sense that it was too little, too late. The loss of revenue and confidence was seen as a natural consequence of the perceived complacency and arrogance of the TV industry across five decades and whatever TV was to do in the future, it would all be a matter of too little, too late. TV was dead!

DEATH

If you Google the phrase 'TV is dead' you will get almost half a million results. This phrase has been circulating since at least the early 1990s. In *'Being Digital'*, Negroponte predicted that the easy storage and distribution of television programming would severely weaken the network schedules and we would all effectively become our own schedulers. To be fair to Negroponte, he did not predict the death of television; he merely predicted that *"the future of television is to stop thinking of television as television"*. Wise words.

At the same time, commentators such as George Gilder began to take this thinking to an illogical extreme. In 1992, in his book *'Life After Television'*[6], Gilder predicted that by the end of the decade (the 1990s, that is), television would be finished as a device and as a business. The impact of greater connectivity and pent up consumer frustration with the limitations of the broadcasting model would be devastating.

This vision saw the death of television as inevitable, rationalized by perceived inherent flaws in its nature, even to the extent of casting TV as a malignant social force to be 'overthrown'. So, Gilder saw television as *"at heart a totalitarian medium"* with its viewers *"sinking into a passive stupor before the tube"*, and *"its overthrow [was to] be a major force for freedom and individuality, culture and morality"* Television had cheapened public tastes and controlled the agenda, for example *'beatifying'* Dr. Martin Luther King at the expense of Richard Nixon.

Such blinkered analysis stemmed from a <u>failure</u> to stop thinking about television as television - the term 'television' almost always denoted the device rather than the platform. The focus of the critique was on the mechanics not the medium itself. Many of the things said to signal television's demise – on demand, the PVR, smartphones, tablets, social media – have become an integral part of the TV ecosystem, adding to and not detracting from the richness of the TV experience.

If, instead of the device, we looked at the future of long-form, narrative-led, audio-visual entertainment, I think there would be a consensus that it will still form an important part of most people's lives in the future, possibly an even more important role than it plays already. In some senses it is irrelevant whether it is broadcast or on demand, viewed on a TV set or a tablet, watched over the airwaves or via fibre optic cable; it is still experienced as 'television'. We cannot continue with the simplistic attitude that, if it isn't viewed on the TV screen from a currently recognised broadcaster, it must be competitive.

The 'death of TV' narrative continued into the new millennium. In 2007, TechCrunch, the digital tabloid, declared *"Why Don't We Just Declare TV Dead and Move On?"* stating that within five years <u>all</u> television content would be delivered via IP with devastating consequences for the broadcast model. Five years later, IP delivery of television remains only a tiny proportion of total TV consumption.

Vint Cerf predicted in 2007 that we were witnessing the end of 'traditional television', which was reaching its *'ipod moment'*. This claim was based on the mistaken assessment that the vast majority of video consumed in the USA was already being consumed away from the TV schedules. That is completely wide of the mark even in 2012; it was a laughable claim for 2007!

Perhaps the best insight into the online world's <u>desire</u> to see the end of television can be found in the pages of Chris Anderson's *'The Long Tail'*[7], where he writes;

> *"TV is not vulgar and prurient and dumb because the people who compose the audience are vulgar and dumb. Television is the way it is simply because people tend to be extremely similar in their vulgar and prurient and dumb interests and wildly different in their refined and aesthetic and noble interests. The Internet has finally allowed us to rise above the mass audience programming requirements of TV"*

This sentiment towards television from the 'always-on' digital fundamentalists is unsurprising. Television is actually 'switch-off' entertainment; we often don't want to go to the trouble of finding the refined and aesthetic and noble at all, and if we do we would rather it be 'curated' than discover it in its raw form. This has often been perceived as time-wasting, dumbing down and passive, and ever since the dawn of broadcasting, experts and commentators have decried TV for these attributes. I'm not quite sure where internet porn, trolling, cyber-bullying or online gambling sit within this analysis (nor Sky Arts, BBC News or HBO drama), but we'll leave that for the moment.

Since its invention, TV has been blamed for rotting brain cells, ruining eyesight, destroying the concept of community, brainwashing, dumbing down, creating obesity, causing criminality (listen to the reports about TV advertising creating the consumer desires that influenced the looting during the 2001 UK riots), shortening attention spans and demolishing society's value systems. Only in the last decade or so has it also been accused of being in terminal decline. Although these seem like opposing opinions – how can something that affects us so strongly be irrelevant or in decline? – I fear that they come from the same source. Visions of TV's imminent end were not based on evidence or objective reasoning. They were polemical – telling us we should always exercise our new power to choose. They disapprove of us making "easy" decisions on what we watch. They are bound up with an agenda which perceives the possibilities afforded by the internet as necessarily trumping the many great attributes of TV and a refusal to understand how a medium which unites people in a common experience of consumption can not only survive but prosper when faced with competition from other digital media which allow for infinite personal choice.

It is not about objective fact or evidence-based reasoning, it is about furthering an agenda, mixed in with a real inability to understand why the masses could possibly value something so *"vulgar and prurient and dumb"*!

TV is not the first medium to have died!

Every new media channel or technological innovation has attracted the revolutionaries, who have proclaimed the death of whatever has gone before. Although many predicted the death of radio, 'inferior' to TV as it can only broadcast in sound, radio redefined its role as a secondary medium that works as a companion device for other activities. It has done so successfully; radio is still much loved, attracts significant audiences (almost a fifth of all media time in the UK is taken up with radio) and still manages to maintain its role as the gateway to new music (it is estimated that around four fifths of the music we hear – even in the i-pod age – emanates from broadcast media, primarily the radio[8]).

An even older medium than radio has been pronounced dead many times; at least four times in my own lifetime. It has been battered and bruised by technological innovation, been disrupted at least once a decade in recent times and has seen its revenue fall victim to the piracy problems that have beset music; yet nowadays very few people consider it to be a deceased, or even mortally wounded industry. That medium is cinema, and it had been considered to be under greater threat from the silver light of the TV screen than even radio had been.

Cinema's history began with the invention of the first cinematic camera – the Kinetograph – which was launched at the 1893 Chicago World Fair. Public performances began in Paris in 1895, pioneered by the Lumiere brothers, and cinema very quickly spread as a commercial proposition. The realism of those early flickering images convinced those nineteenth century Parisian audiences that the approaching train on screen really would crash into them physically; similar 'mind tricks' involving a suspension of disbelief that has continued through the decades, from *'2001'* to *'Star Wars'* all the way through to *'Avatar in 3D'*.

In the first half of the twentieth century, cinema developed technologically, through the introduction of animation in 1917, sound in 1927 and colour in 1935. The swathe of 3-D movies in the 1950s was a direct response to the threat from TV, providing an experience TV could not match (then, at any rate). Movie theatres also became more grandiose and well-equipped, ensuring that the cinema experience provided maximum value.

Once television reached mass penetration in the 1950s, the predictions of cinema's death multiplied. The worry was that the same, rich audio-visual storytelling that people had to leave their homes to experience and had to pay to see would be critically affected when it was beamed into people's own living rooms, for free.

Cinema's death has been predicted many times since, almost always in response to technological innovation. The mass take-up of colour television in the 1970s; the sudden rise of the VCR market in the early 1980s; the roll-out of cheap DVD players in the early 2000s and the digital revolution, leading to the easy access and storage of on demand cinematic content, have all been cited as the final nail in cinema's coffin.

In 1983, as the VCR reached mainstream status, Jack Valenti, the head of the Motion Picture Association, likened its impact to the Boston Strangler. The DVD, PVR and internet have all been accused of killing cinema. Even the humble TV remote control device has been cited as a major threat. As Peter Greenaway, the acclaimed British director, stated just five years ago;

> *"Cinema's death date was 31 September 1983, when the remote-control zapper was introduced to the living room, because now cinema has to be interactive, multi-media art,"*

In '*Being Digital*' Nicholas Negroponte wrote about the threat cinema faced from the ubiquity, speed, convenience and lower costs of entry that rich audio-visual media could enjoy from the digital revolution. Why would people pay top dollar to see the same content outside the comfort of their own homes? In fact, the argument sounded eerily familiar to the previous obituaries for cinema, following the mass adoption of TV sets into American homes.

Of course, cinema is by no means dead, although the cinematic experience has evolved fundamentally to respond to the perceived threats. Cinema has benefitted enormously from the digital revolution, transforming production costs and capabilities, opening up distribution channels (so, for example, movies that would have taken almost a year to transfer from the US to Europe can now be premiered simultaneously) and creating promotional opportunities that the analogue film industry could never have imagined.

All of which has transformed the movie industry's business

model. Until the first television deals, the film industry was totally reliant on the revenues from theatrical release. These revenues were considered substantial in their day; by 1946 around 60 million Americans were visiting the movie theatres on a weekly basis and in the UK at least four out of every five adults visited at least once a year. Box office receipts in the USA hit a high of $1.7bn in 1946, but the advent of television undoubtedly contributed to a long-term fall in box office admissions. Fortunately, new revenue streams were emerging from the technologies that were meant to destroy cinema.

Television, as well as being the main cause of the film industry's woes, was also the precursor of the more diverse business model we see today. In 2008, global cinema revenues were estimated at $65bn, of which only around a quarter can be attributed to cinema admissions. Two thirds are taken up with DVD sales and television rights. In fact, TV rights have been more valuable than box office receipts since the mid 1980s, and are still increasing. Overall, it is the non-theatrical revenues that have been responsible for an estimated doubling of movie industry revenues since 2000. Hardly the death throes of an ailing medium.

Edward Jay Epstein, in his book 'The Hollywood Economist[9]' describes the changes in the cinema business model as follows;

> *"There was a time, around the middle of the twentieth century, when the box office numbers that were reported in newspapers were relevant to the fortunes of Hollywood: studios owned the major theater chains and made virtually all their profits from their theater ticket sales. This was a time before television sets became ubiquitous in American homes, and before movies could be made digital for DVDs and downloads.*
>
> *Today, Hollywood studios are in a very different business: creating rights that can be licensed, sold, and*

leveraged over different platforms, including television, DVD, and video games. Box office sales no longer play nearly as important a role. And yet newspapers, as if unable to comprehend the change, continue to breathlessly report these numbers every week, often on their front pages. With few exceptions, this anachronistic ritual is what passes for reporting on the business of Hollywood."

Just as we are seeing in the TV market, the media concentration on the headline figures masks the real engine of the industry in terms of both revenues and audience behaviour.

We have learned a great deal about death, and how it is reported, through the experiences of radio and cinema (never mind the death of print, newspapers, books, writing...even the internet itself) but that has not prevented similar untruths, misperceptions and poor analysis contributing to the perceived death of the medium that had been considered a media-killer itself just a few decades earlier. I mean, of course, television!

PART TWO

NOW

PERFORMANCE

In this chapter, we look at television's current performance, including the myths surrounding the evidence of its predicted demise.

Recent rapid changes in the way we consume and access media have led some commentators to conclude "TV is dead". If we examine the various reasons justifying these forecasts, with a more sophisticated eye, a complex landscape emerges. My view, supported by empirical research, is that, whatever impact these changes are having, however the landscape is being reformed, one thing is clear - television is performing better than ever before. TV is not dead, it's different and the way we use it will continue to evolve.

Here are the six recurring myths upon which the 'TV is dead' narrative is based:-

1. People don't watch television any more
2. Fragmentation is inevitable...and bad!
3. Audiences will free themselves from the schedules and 'migrate' to other alternatives
4. PVRs will destroy the thirty second spot

5. Advertising is dead anyway
6. The death of content

Let's analyse television's current performance through that lens.

1. **People don't watch TV any more**

The most common myth about TV's decline is that '*nobody watches television any more*'. But evidence from the wider world of industry ratings systems and other trustworthy research sources (not a sample of internet panel members asked how many hours a day/days a week they watch TV) is clear. Television viewing has never been higher.

Since 2007, almost every month has produced a new record for the amount of time we spend watching TV. In 2011, record hours of viewing were reported[10] in the USA, the UK, Canada, Germany, France, Austria, Belgium, Italy, Ireland, Portugal, Holland, Spain and Australia, to name just a few, and 2012 has continued this trend. These figures are only the reported figures, covering viewing of broadcast content via the television screen. None of the new forms of viewing, supposedly driving viewers away from the TV screen, are included, nor is recorded content beyond seven days. Nevertheless, TV viewing just keeps on rising. How can this be?

We have to revisit Negroponte's assertion that we need to stop thinking of television as television. Not only has technology found new ways for us to access TV, but the content itself is far more engaging. We are watching on bigger and better screens. Production values are improving all of the time. We can store, share, enhance and personalise the TV we love like never before. There has never been so much quality content to watch, at all times of the day and night. That is why TV viewing continues to rise.

Performance

The rise is across virtually all demographics, including children and younger or more affluent adults. These audiences may not be watching as much as other audiences, such as older adults, but we know that much of the shortfall will be taken up by TV viewing via other platforms.

The UK provides a good snapshot of what has happened globally. During the 1970s, the average individual watched just over three hours of television per day. This rose by around half an hour per day during the 1980s and 1990s, mainly due to extended broadcasting hours. It then stayed flat, at around three hours 40 minutes a day, until 2007. Since then, average total TV viewing has increased by over 20% and now stands at an average four hours and 20 minutes every day; not just the highest viewing levels on record, but a record rate of growth. To put this into perspective, the increased amount of TV viewing alone since 2007 is greater than the total amount of time British people spend on all social network sites every year.

TV's dramatic rise in viewing since 2007 has coincided with the growth of the digital and online (they are not the same thing) technologies that were supposed to have killed it.

One of the main reasons for the perception of TV viewing decline, despite so much evidence to the contrary, is that people often concentrate on the headline figures based on the performance of the big programmes and channels. In the UK, ITV witnessed a dramatic decline. To fall from 39% share to just over 20% - to lose almost half of your audience from a seemingly impregnable position in less than a decade – shattered confidence in the power of the broadcast networks. The even steeper decline of individual programmes, especially in the all-important peak-time slots, created a belief that even ratings bankers, such as '*Coronation Street*' or '*Emmerdale*', were not long for this world.

Generally, the top-rating shows in the UK achieve average audiences of 10-12 million viewers. Just ten years ago, the top programmes were averaging anything from 14 to 19 million and the number of 10 million+ shows was around twice the number we see today. These are the figures that we see reported and the assumption has been that this is reflective of programme audiences overall.

If we add in on demand viewing, delayed timeshift and same-week repeats, the two sets of figures would look remarkably close. The big hits now create their value, just like cinema blockbusters, through a variety of different channels across a period of time, rather than just on their release date. The point to remember is, globally, television viewing (just via scheduled programmes on the main TV set) is the highest in television's 75 year history. That picture is in sharp contrast to what the media headlines would suggest. But that is the problem with headlines; people often don't look beyond them.

2. **Fragmentation is inevitable...and bad for TV**

Fragmentation is inevitable. If new channels are introduced then either they will achieve zero take-up (which never happens; even the most obscure programmes and channels will get some viewers) or they automatically create further fragmentation of the existing audience.

In the case of television, the number of channels received by the average household has risen from just under four in the mid 1980s to well over a hundred in 2011. All but a tiny proportion of the population have ready access to at least 30 channels and more than half of them have access to hundreds via cable or DTH satellite. Many of those channels often get zero ratings for individual programmes but they still get an audience and, combined, it is a sizeable one. More mainstream channels such as Sky Atlantic (launched in early 2011) continue to drive audiences further afield.

So, fragmentation is happening, but all the signs are that fragmentation will have less of an impact on TV ad revenues in the UK than has been predicted because;

1. The advertising revenue premium gained for the programmes that do reach large audiences, or that act like social glue for a specific subset of the audience, is increasing and goes a significant way towards countering the fragmentation trend.

2. Although the overnight figures (the ones that are usually reported) are significantly lower than for similar programmes in previous decades, once we add in the accumulated audience from on demand, timeshift and repeat channels across the rest of the week, both see significant gains, often bringing them up to 'pre-fragmentation' levels. The bigger the programme, the greater this effect in general. Some mainstream programmes now achieve more than half their audience outside the original transmission.

3. The fragmentation of channel viewing is not being matched by fragmentation of broadcaster power. The consolidation of sales points in the UK, together with the success of the big broadcasters' digital channels and on demand services, has created a more cohesive broadcast market than we've had since the launch of Sky Digital.

4. The ability for TV to create social glue will be explored further in later chapters but, the bottom line is that social media and online technologies are attracting sizeable audience communities around favourite programming and even attracting viewers back to the live broadcasts so they can share the experience in real time. This is especially true for the peak-time favourites.

It could be argued that the fragmentation trend is reversing. Not only has there been a further concentration of broadcaster power in the advertising sales market (as mentioned above) but the broadcaster digital portfolios have been relatively successful in fighting off the pressures of fragmentation, mainly through expansion and better branding. This may help explain why, in 2010, the number of broadcast channels in the UK actually fell for the first time since 2004.

We should also refrain from the 'fragmentation is bad' hypothesis. While marketers would love all TV programmes to reach several million viewers, that would mean also accepting a decline in targeting efficiency. In the early 1980s, for example, only one TV programme delivered an above average proportion of young adults (Channel 4's '*The Tube*'), making targeting inefficient and difficult to apply. Nowadays, we have whole channels dedicated to specialist audiences or niche demographic groupings. In fact, TV's increasing ability to efficiently reach niche groups of viewers is one of the main reasons cited for its improving cost-effectiveness.

3. On demand: the great 'migration'

One of the phrases that I have found most misleading over recent years has been the concept of audiences 'migrating' to other alternatives. The concept of migration is of a long-term movement from one location to another. This doesn't tie in with the irregularity of visits and short dwell times that the vast majority of 'migrants' devote to these new devices and platforms.

In the UK, TV on demand has been an unqualified success. Since the launch of the BBC i-player, it has been enjoying strong growth in audiences and revenues. Four out of every five broadband users watch TV online regularly and viewing is increasing by around 25% a year.

Let's put it into perspective, though; the combined share of our TV time taken up by these 'competitive' on demand experiences is less than 3%. Even when we add in timeshifted viewing of broadcast content via the PVR, the share of total TV viewing taken up by live viewing to the broadcast schedules is well over 90%, and most experts now agree that it is likely to remain the dominant model for the foreseeable future. So, this is hardly a migrating audience, more one composed of casual travellers and day trippers.

As well as mainly confining their experiences of these new TV destinations to short, irregular visits, those trippers are also booking return tickets back to the schedules. Often, the motivation for catching a programme on demand is because people missed the show when it was broadcast at the scheduled time, and have elected to catch up later with every intention to go back to the schedules for the next episode. The availability of on demand catch-up keeps viewers loyal to a series. If this is migration, it is hardly producing a settled population.

This all suggests the desire to 'break free' from the tyranny of the network schedules has been severely overstated. The argument runs that broadcast schedules were necessary in the days when people had few alternatives but, when the opportunity presents itself, we would each become our own scheduler, maximising our enjoyment from the increasing array of programmes and channels that online has made available to us in our own time and at our own convenience.

Catching up on the programmes that have recently been scheduled live is the main driver of on demand viewing most of the time, and it appears to be growing in influence[11]. The desire to find alternatives to the network schedules and to discover new programmes is being squeezed out by the plethora of opportunities to catch up on already established favourites. But, the question needs to be asked, why do scheduled programmes from the favourite channels form the vast bulk of time shift or on demand in the first place? Maybe, just like theatrical release for a movie guarantees it must be better than 'straight to DVD', the channel schedules provide the same guarantees for TV viewers.

There is also a growing perception that, for most viewers most of the time, the schedules are not a ball and chain at all, but provide a 'curated' experience from a trusted editor timed to fit in with the structure of individuals' and family lifestyles in the best possible way. Therefore, rather than breaking free of the restrictions of the schedule, viewers appear to be using these new opportunities to maintain the schedule's hold, by increasing their loyalty to programmes and series, which they will then return to the schedules to view next time. It is additive, not competitive.

The desire to break free from the constraints of the schedule was assumed to be the driving force behind the rapid adoption of video cassette recorders (VCRs) in the 1970's but remarkably, just as with PVRs a quarter of a century later, they achieved around 15% of viewing within the homes that used them – despite being less convenient and easy to use than PVRs. It appears that there is a fairly fixed ceiling in how far away from the schedules most people are prepared to stray.

There are three main influences behind the resilience of the schedule.

1. The majority of TV viewing is shared (a point I'll return to later) and this appears to be a growing phenomenon. The need to share acts as a powerful driver towards watching live; most shared viewing is based around the schedules whereas on demand tends to be a more individual experience.
2. The ephemeral nature of television is a powerful driver – what's on <u>now </u>will always be more important than what

we can watch at our own leisure (as anybody who has found themselves watching a movie on TV when they have the DVD just a couple of metres away will testify).

3. Schedulers <u>know</u> their viewers and know what programmes work best in different dayparts and even days of the week (e.g. gritty dramas on a Monday night but more gentle fare on Sunday evenings). It is no coincidence that many families continue to schedule other family activities (mealtimes, bedtimes, going out) based on particular schedule structures such as news, soaps and major programming events.

The power of the schedule is a focus of Deloitte's 2012 media predictions; as Jolyon Barker, who heads Deloitte's Global Technology, Media and Communications practice says;

"Technology has not shattered the TV schedule; it has made it more resilient, more flexible"

4. The PVR – the death of the 30 second spot?

When the first digital TV recorder launched in 1999 there was little doubt that the technology would be a success, but the overriding narrative was not about how it would enhance and increase TV viewing but about how it would destroy the 30 second spot. Why, the experts reasoned, would people sit and watch the commercials they had always professed to find so irritating when they could arrange to watch all their favourite programmes in time shift mode and then fast forward through the ads in just a few seconds?

In the UK, mainly due to the success of BSkyB's marketing of Sky Plus, the number of homes owning a PVR has grown to

50% in less than a decade. During that time, not only has total viewing grown consistently and significantly but the viewing of TV commercials *at normal speed* has grown even faster. When PVRs first came to market, the average UK viewer watched on average 38 commercials in full every day. Now they watch almost 48 – a full 10 commercials a day more exposure for every man woman and child in the nation, despite half of them having the ability to fast forward those ads out of their lives. How can this be?

Part of the reason for the growth in commercial impacts is that we simply watch more commercials; a greater proportion of UK viewing is going to non-PSB (public service broadcaster) commercial channels, and those channels are allowed more advertising minutage than the three commercial PSBs. So, consumers, who are watching more TV in the first place, are exposed to more advertising per hour viewed.

The question arises - shouldn't this give PVR owners more reason to avoid the ads? It hasn't worked out that way; in fact, PVR owners watch more ads at normal speed than those without the technology. They watch more ads at normal speed than before they got the technology themselves. Industry data shows how consumers have totally confounded the expert predictions.

The first thing to note is that PVR ownership means that people immediately view a lot more television, for one simple reason; no matter what time you turn the TV on, there will always be something you want to watch. The average increase in viewing upon adoption of a PVR is 15% (which is consistent across countries and research sources) but the proportion of total viewing devoted to time shifted viewing of recorded programmes is...around 15%! In other words, PVR time shift viewing is pretty much all additional. It appears that people are primarily using the PVR to find something to watch at times when they would not normally be viewing TV, reflected in the fact that peak-time programmes are often played back in off peak time slots. We were so busy looking for the PVR's destructive qualities, we failed to spot that consumers were buying them for much more positive reasons; to be able to watch more of the TV they love, at times that suited them.

An interesting side note to this point; if consumers were so focussed on skipping the ads, time shifting levels for the commercial channels would be much higher than those for the ad-free BBC. It would make much more sense to watch BBC programmes live, ad-free, and mainly record the commercial channels for later fast-forwarding. In fact, time shift for BBC channels is actually higher (7.5%) than for the commercial channels (6.9%)[12]. This demonstrates beyond doubt that ad avoidance is not a major driver of PVR adoption or use.

Even when they do record programmes onto their PVRs, those consumers are not lost to the advertiser. On average, PVR owners watch around a third of time shifted commercials *at normal speed*. Often it is because they forget they are watching in time shift mode (a big chunk of playback is via live pause). There is also plenty of evidence that they will often stop fast forwarding in order to share a particularly entertaining or relevant commercial – a valuable advertising opportunity if ever there was one.

What this all adds up to is a net increase in the viewing of TV ads *at normal speed* of two to three percentage points compared to before the PVR was installed.

How PVRs Affect Viewing to TV Ads

Average Increase in Household Viewing to TV	**+15%**
Average % Household Viewing Time shifted	**15%**
Average % Household Viewing Watched Live (includes second sets)	**85%**
% Time Shifted Ad Breaks Fast Forwarded (average at 12x)	**65%**

% Time Shifted Ad Breaks Watched at Normal Speed	35%
Average Increase in Household Ad Exposures	+23%

Far from destroying the 30 second spot, PVRs appear to be increasing our exposure to TV advertising across the board (there is also strong evidence that fast forwarding increases the impact of sponsorship bumpers, as they are used as signposts). This is true of just about every developed TV market and is consistent across every research tool used to measure it (e.g. industry currencies, server-based data such as BSkyB's Skyview panel, ethnographic observation); just as long as it isn't self-reported. People aren't very good at remembering this stuff for themselves, especially after the event.

One of the reasons these astonishing stats are met with almost universal disbelief by marketers is that they seem so totally at odds with their own experience. Many industry professionals will tell you that they hardly ever watch ads any more (a strange statement from somebody working in an advertising industry) but they are either very atypical, or they are wrong.

An insight into this phenomenon comes from a study initiated by Professor Paddy Barwise, Emeritus Professor of Marketing at London Business School. Professor Barwise has developed a great understanding of how TV audiences behave, not least through his work with Andrew Ehrenberg in the 1980s, and he sensed some of these apocalyptic predictions were not rooted in real world evidence, so he decided to collect the evidence for himself. Rather than asking people, however, he observed them. Working with ethnography specialists ACB during the summer of 2006, he installed cameras in a representative sample of homes (all early adopters of PVR technology) and watched their behaviour, which was also analysed by a team of trained ethnographers. He found that, on average, they watched around 15% of their viewing to time shifted material, and only fast forwarded the ads around 70%

of the time; real life behaviour exactly matching what BARB and Skyview are now telling us.

The most interesting outcome of this research, though, was that when the respondents were questioned about their PVR behaviour, even though they knew they had been observed, they insisted they watched around three times more in time shift than they actually had. Because that behaviour was active and required attention, they completely overstated its presence in their viewing repertoire.

In conclusion, PVRs show no indication of destroying the TV advertising model; indeed they appear to be strengthening it. People don't adopt PVRs to avoid ads, but to improve their viewing experience and watch more programmes. The fact that not only does that expose them to more advertising, but that they are more likely to be engaged with the surrounding programmes, shows how even the most 'disruptive' technologies can prove to be a blessing in disguise.

What all of this means is a thriving spot advertising market. Viewing to TV commercials is at astonishing levels. In the past five years alone, when the TV spot has been under so much pressure, the time spent watching UK commercials *at normal speed* increased by almost 25% and reached record levels for every main trading audience. Even the digital natives were watching more spots, even if they were also Facebooking their friends at the same time; but then, as we shall see, that is much more of an opportunity than a threat to TV.

5. Advertising is dead – or is it?

The death of advertising has been predicted for some time. Take this quote from a magazine article;

"Advertisements are now so numerous that they are very

negligently perused, and it is therefore become necessary to gain attention by magnificence of promises, and by eloquence sometimes sublime and sometimes pathetic."

It comes from a magazine called The Idler and it is the first reported comment on ad avoidance. It was written by celebrated London chronicler Dr Samuel Johnson in 1759.

Dr. Johnson was referring to the explosion of newspapers, magazines (like his own), billboards and other forms of print media that characterised mid 18th century society, but it is a refrain that has been heard regularly, although not quite so magnificently, ever since.

We have always expected audiences to try to avoid advertising; maybe that is due to a sense of low self-esteem on those who create it. Advertising is everywhere and, just as with the linear schedule, its death has been widely presumed.

There are many reasons why the whole principle of advertising avoidance, particularly relating to television, has been consistently overstated. Viewers have always had ways of avoiding the advertising, which is why advertisers keep pulling the levers of creativity to keep them there. Despite Dr. Johnson's fears, it still appears to be working...two and a half centuries later.

The 'death of advertising' lobby points to the recent declines in TV revenues (all of which were clawed back in 2010) as fundamental evidence that the world is moving to an ad-free alternative. They point to the increase in marketing activities that seek to by-pass the paid-for media in favour of 'owned' media (e.g. the company's website, free magazine or branded channel) or 'earned' media (such as viral views on YouTube or social media conversations). They talk about 'the end of the age of interruption' when most of our brand conversations are created through interruptive advertising in the first place. Given the resilience of advertising revenues, though, as well as the desire for brands to reach as many consumers as possible, this does not suggest the death of advertising any more than Dr. Johnson's comments from two and a half centuries ago.

Advertising revenues aren't dying, they are shifting, recently from display to classified (through the phenomenal growth of the search advertising market) and from brand building to response. So far, that shift has not radically impacted on TV's revenue base, contrary to expectation. Plus, there is plenty of evidence (explored in more detail in the *'Payback'* chapter later) that the tide is turning, and the role of display advertising is becoming more valued.

6. The death of 'content'?

The final, often vaguely outlined threat to television's future, I will refer to as the Apocalypse View. Content will be free and easily accessible, its value will decline exponentially, advertising will be instantly avoidable and we will consume content we generate ourselves. All of the Apocalyptics' predictions are vague on what this replacement content will look like, but all tend to agree that the 'traditional' media channels will crumble in its wake.

Yet, the more choice that is available, the more people are returning to the 'trusted editors' to help manage it. Nowhere is this truer than for television. Despite ever-increasing alternative ways to find any video content ever made, scheduled TV still forms the vast bulk of our audio-visual entertainment, a massively-growing market in itself. Consumers are still flocking to the schedules and channel brands and EPGs are their main navigational tools.

But TV's real power doesn't lie in the technology, it lies in the content. TV offers high quality, curated, rich media storytelling entertainment and people will always flock to that. So will advertisers. Given that brands will always be with us and the financial impact of branding is becoming better understood over time, all the evidence suggests TV will be able to offer scale, impact, engagement and effectiveness to justify their investment. Add in the willingness of consumers to pay for that same high quality, narrative-led entertainment content (let's just call it TV for short), then we should follow the money.

To the Apocalyptics, I say people will always be prepared to pay for better access and better content, but if they can get it for free with some advertising included, they are remarkably receptive to that option. Unless a cost-free, ad-free, hassle-free alternative is easily available, consumers will follow the content, and where consumers congregate, especially if they are engaged and receptive, then brands will seek them out. Some forms of content may have become commoditised, but TV is performing better than all expectations. There are good reasons for that.

The following chapters will look at <u>why</u> TV is thriving – as an entertainment medium and an advertising channel. It has something to do with technology, but it is driven by people. We will look at concepts like emotion, engagement, storytelling, sharing and trust; concepts that have been a vital part of the human condition since the dawn of time and have influenced television's renaissance far more than any technology.

Before we do, we need to ask the question, how could we have ignored all of this evidence when the 'TV is dead' narrative took hold? Why did it take hold?

WHY?

I grew up loving TV, so there was an emotional element to my resistance to the 'TV is dead' orthodoxy - but then emotions drive most of what we do. But I also resisted it as a researcher; I had made my career through commissioning, analysing and presenting research on media behaviour, and none of that data was telling me TV was dying, at least not as far as audiences were concerned. At Thinkbox, every day brought some new statement in the press saying why *x, y* or *z* would be replacing TV soon and we would have to respond with hard evidence to back up our counter claims. In five years, I never saw one piece of convincing evidence to support the predictions. There was plenty of data flying around... but I would challenge anybody to find a report from the last five years that offers convincing evidence based on actual consumer behaviour to demonstrate any signs of TV's deterioration at all!

The problems of dodgy data

Where did the data come from that pointed to the collapse in viewing, the fast forwarding of all advertising breaks and the migration of mass audiences to niche viewing services? Surely there must have been some data to back these proclamations up...

mustn't there?

There have been several predictions over the last decade that did refer to research findings to argue that PVR owners were fast forwarding the vast majority of the ad breaks or that people were spending more time on the internet than they were watching television. The reason why such claims made it to the front pages of an increasingly credulous trade press was that, on the surface, they appeared to be based on strong evidence. To the professional media researcher, they contained a multitude of errors in the way the data was collated, analysed and interpreted.

The fear of the PVR wiping out the audience to spot advertising dominated media and marketing thinking in the first half of the last decade, mainly based on research among early adopters. When those early adopters were asked how often they used their PVR and how often they fast forwarded through the ads, they said 'most of the time' in answer to both questions. Now we know that such self-reporting was severely overstated, and it was an unrepresentative, early adopter sample, but such early impressions did help to fuel a perception that, despite massive evidence to the contrary, has never been completely dispelled.

Similarly, the decline in TV viewing was assumed because naive researchers asked a self-selecting online panel (usually skewing toward the heaviest online users) approximately how many hours a day they watched television and how many hours a day they spent online and then – *quel surprise* – found that claimed online time eclipsed TV viewing. Microsoft and the European Interactive Advertising Association (EIAA) have put out this kind of research, accompanied by the predictable headline for several years on the run now.

When questioned about the methodological constraints of just collecting data from a self-selecting and unrepresentative group of people, and using simplistic 'how many hours a day on average' type questions to arrive at such far-reaching conclusions, the digerati will turn around and question how the TV industry can rely on a small 5,000 sample to provide the currency for an entire £3 billion+ industry? Surely the bigger sample sizes from online panels make their research much more valid and accurate. Actually, no.

BARB, and most of the other ratings currencies, are based on household samples (so the 5,000+ BARB panel equates to 12,000+ people) and are selected specifically so they represent the overall population against all of the factors that we know influence TV viewing. They are monitored and checked continually and viewing behaviour is tracked passively so it doesn't rely on either memory or reporting accuracy. Having sat on more BARB committees than I care to remember, I can testify to the constant quality control and evaluation of the service as well as the numerous systems and processes in place to pick up any anomalies or errors in the database. BARB, like any currency, needs to be trusted, reliable and accurate - much more so than asking a bunch of random people a list of unanswerable of questions, set in a biased context and totally excluding around a quarter of the population who never or rarely go online.

BARB has recently been validated by Touchpoints[13], and most other viewing research that monitors viewing choices continually rather than just asking the question in a vacuum; all three Touchpoints surveys so far have shown a 99%+ correlation with BARB-reported total TV viewing patterns. However, there are other ways of looking at how people view television, the best being not to ask them but to watch them instead.

I have already mentioned the ethnographic work undertaken by ACB in the UK. The Council for Research Excellence (CRE) in the USA conducted their own ethnographic study into how people use screens in real life and came up with very similar conclusions. The new activities and behaviours had been massively over-reported and real people still relied on television as their primary source of entertainment much more than had been expected.

The 2008 CRE Video Mapping Study sent trained ethnographers to follow a representative sample of 476 US consumers for every waking minute over two days each, to simultaneously measure their media exposure, their life activities and the locations where they spent their day. It was conducted because self-reporting methodologies were failing to keep up with the complexities of modern media usage. As such, almost 1,000 days were observed and recorded in detail, to show what really happens in American consumers' lives.

The study showed that average screen time is very consistent across all ages and social groups, but different demographics have different screen behaviours. However, across all demographics, watching live programming on the TV screen totally dominated screen time; even among 18-24s, 41% of their screen time was live TV, followed by online communication at 13%. However, when it came to watching video-based entertainment content, TV content accounted for 99% of all viewing. For 18-24 year olds, the figure was 98%. Given that audio visual is the communications medium of choice for most people and the most effective advertising format, this suggests it is going to be difficult to break TV's domination in this area.

Observational techniques like this also reinforce the accuracy of the ratings systems. They all point to the flaws in simply asking people – any group of people who happen to volunteer for the job (usually for money) – about their increasingly complex media usage. They demonstrate that one needs to measure these things accurately, sensitively and consistently. Interestingly, in the UK, the online industry is desperately trying to provide a BARB-like measurement service for online display (UKOM), for which the current range of analytics solutions are completely unsuitable.

The problem is that if people wish to believe flawed and inaccurate data to suit their own ends, it is very difficult to persuade them otherwise. The digital industry and their cheerleaders have needed a lot of convincing. It is almost certainly because they have been using the wrong data.

The problems of binary thinking

The whole basis of digital technology is zeros and ones forming a binary code around which complex pieces of content can be created. A great deal of digital analysis appears to be based on a form of binary thinking – either/or, one thing replacing another, winners and losers, dead or alive. It assumes for one thing to succeed it must be at the expense of another.

Binary thinking is far too simplistic to apply to people. Cognitive psychology, neuroscience and behavioural economics have all demonstrated the complexity, irrationality and unpredictability of human behaviour.

The beguiling simplicity, immediacy and (seemingly) comprehensive nature of the data that the digital universe provides convinced many of us that it had all of the answers. It had been a triumph for analytics over insight, digital over analogue, rational over emotional. So, in this context, it is easy to see how the digerati got it wrong. The analytics gave a misleading sense of untrammelled growth for digital, and the experts made the assumption that it must all have been at the expense of the existing 'competitors' – a zero sum game. That is binary thinking. The fact that analytics didn't provide the insight into where that growth came from, nor how real or sustainable it was, how it related to primal human needs, or how those needs were currently being satisfied was largely ignored.

Binary thinking has created an *us versus them* mentality, digital versus analogue, assuming that it would be the digital experiences that inevitably come out on top. But it has produced a form of digital fundamentalism that temporarily closed the industry's eyes to the real behavioural changes that have been taking place below the surface.

As we have already seen, television hasn't died; it didn't even fall ill, but its resilience is not based on new technology, it is driven by fundamental human needs and behaviours that are actually analogue in nature. Unless those experts who confidently predict our media futures start to understand this, they will be condemned to walk the earth, constantly repeating the same mistakes and miscalculations, over and over. To quote the philosopher, George Santanaya;

> *"Progress, far from consisting in change, depends on retentiveness .... those who cannot remember the past are condemned to repeat it"*

ENGAGEMENT

'Engagement' must be one of the most over-used and under-defined words ever employed in a business context. It has become a mantra within media, ever since we moved out of the 'Age of Interruption' and into the 'Age of Engagement'. According to more than one media agency, we have now moved into the 'Age of Dialogue' - we left the Age of Engagement without fully understanding what it is.

Until recently, media planning had been based on eyeballs rather than engagement. TV programmes were generally judged by the ratings and most advertising campaigns were judged according to the reach and frequency (how many people saw the ads and how many times). Then, along came engagement. From the late 1990's onwards, consumer engagement became the Holy Grail. The problem is that there has been little insight into what it means or how it can best be defined.

In 2006, more than twenty of the best minds in the American advertising industry came together under the auspices of the Advertising Research Foundation (ARF) to arrive at a consensus definition of 'engagement'. The committee, which included senior representatives from the major advertising agencies and biggest-spending advertisers (including Procter & Gamble and General Motors), agreed on the following;

> ***"Engagement is turning on a prospect to a brand idea, enhanced by the surrounding context"***

The ARF definition describes the three main attributes of engagement; the <u>audience</u> (or prospect); the <u>content</u> (or brand idea); and the <u>context</u>. It gives no indication of where engagement stems from or how it can best be achieved or even measured. To be fair, the ARF has since produced a series of thought pieces and evidence-gathering exercises to evaluate engagement-related measures based on facial expression, emotional metrics, neuroscience, online 'buzz', visual projective techniques and implicit attitude testing. Meanwhile, many campaigns are assessed on a range of 'engagement metrics', generally either online behavioural measures or attitudinal measures such as top-of-mind recall, purchase intent or brand preference.

The reason there has been so much focus on engagement as a concept, for both media owners and advertisers alike, is because however it is measured, it generally indicates success. It has also coincided with the rapid expansion of behavioural data now available from online. To an extent, this has created a sense that 'engagement' is primarily an online phenomenon. There has been a whole industry built up around 'engagement metrics' and, over the last decade or so, advertising engagement has been measured via click-through rates, dwell times, unique users, frequency of visit, recency of visit, 'bounce' measures, sales through site, provision of personal information and social media 'likes'[14]. There is no room in this analysis for the role of offline media which, for many years, tended to get overlooked within this debate.

More recently, engagement has been explored in a wider context. Organisations like the ARF and many of the more progressive market research agencies are beginning to understand that 'engagement' is not necessarily an active behavioural measure and that traditional brand tracking metrics fail to grasp its true nature or contribution to a brand's success. The role of emotion and sub-conscious processing are becoming better understood as new techniques have been appropriated from neuroscience and cognitive psychology. This chimes with our improved understanding of how TV works on the brain, which I'll investigate further in the next chapter, and shows much stronger relationships between these measures of engagement and behavioural outcomes such as usage, purchase, interacting and sharing. In other words, engagement – as we're beginning to understand it - makes money.

Engagement is <u>not</u> attention. Attention is about cognitive, focussed, analytical thought, whereas neuroscientists define engagement as *"a sense of immersion, brought about by feelings of personal relevance"*[1]. Engagement is almost the opposite of attention; emotional rather than rational; implicit, rather than explicit. It is when we have our critical barriers down and just take things in. In this context, engagement is by far the most powerful advertising state. Work conducted by a number of leading neuromarketing experts[15] has shown it is engagement that leads to long-term memory processing, and ultimately to purchase, far more effectively than attention. The behavioural measures of engagement are actually measures of its outcome; people will click through, interact or share if they have been engaged by an advertising communication in the first place. We have been confusing cause and effect.

This ties in very closely with Robert Heath's theory of low involvement processing[16]. Using hundreds of real-life case histories, Heath has shown that advertising eliciting low levels of attention or cognitive processing generally works more

1   As defined by Professor Richard Silberstein, based on his neuroscientific analysis

powerfully than that which gets us to think about and analyse the 'message'. It is the strength of the emotional connections that makes low involvement advertising work so well; it literally gets into our brains from under our cognitive radar. It makes us think we know for ourselves the product benefits and associations that the advertising communicates. Television creates those conditions particularly well, largely because of the content (long-form, rich, audio-visual 'storytelling') and the context (relaxed, receptive, with friends and family).

A case in point is the campaign for Virgin Trains in 2006, featuring digital inserts of actors from classic movies (*'The Railway Children', 'Strangers on a Train', 'Some Like It Hot')* cut into scenes from modern Virgin train journeys. The ad did not perform well in pre-testing - consumers reported that the message was confused and it wasn't 'aspirational' enough (an example of the way consumers often rationalise their own perceptions through marketing jargon). However, implicit testing techniques found that those same consumers held much more positive beliefs about the Virgin Trains service and quality of the experience after watching the commercial. They believed the carriages were modern and comfortable and the journey was smooth and relaxing. They didn't associate these beliefs with the ad itself, because it had worked on them implicitly; they assumed their improved perceptions were just things that they *"knew for themselves"* – surely the real 'Hidden Power of Advertising' according to Heath's acclaimed text on the subject[17]. (Needless to say, the Virgin Trains campaign went on to yield more than £4 in increased sales for every £1 invested).

The impact of campaigns like Virgin Trains', and an increasing number of others, is not through attention. In fact, it is often despite attention. Engagement works in a very different way, as most practitioners in this field now recognise.

When I joined Thinkbox in 2006, my first project was the Engagement Study[18], an ethnographic study watching what happens when viewers watch the commercial breaks. A total of 22 families (comprising 74 individuals) were recruited and cameras were connected to their TV sets. Their viewing was recorded and monitored over six weeks, providing over 17,000 observations of people watching commercials in their natural surroundings, an unprecedented insight into what really goes on when people are watching TV ads. It didn't provide us with any new definitions of engagement...but it showed us what it really looks like.

We filmed people singing along, clapping, dancing, whistling and regularly joining in with the ads in a very playful way. They played 'guess the ad' games with each other, pointed out interesting (and engaging) aspects of the ads, talked about their own experiences of the brand or discussed the ad's 'story'. They were often involved, immersed, and definitely engaged!

Engagement is strongly linked to emotion, which has been identified as a fundamental driver of human behaviour. It is heightened by shared experience, and as we shall see in later chapters, television viewing is becoming an increasingly shared experience. It is fuelled by trust and fame and talkability, all of which we will cover later. Television creates a unique sense of engagement between the audience and the content, within a positive surrounding context. When people are immersed in TV content, as they often are, then the advertising can do what it does best.

We'll return to the definition of engagement later, but I want to stay with the ARF's combination of <u>audience</u>, <u>content</u> and <u>context</u> to demonstrate its importance to television, and how it begins to explain TV's enduring hold on consumers in even the world's most advanced digital market-places.

AUDIENCE

I love studying people; they are far more interesting, complex and unpredictable than technology, markets and economies. I have never yet heard of an economic theory that could explain the world around us as well as Stanley Milgram's 1961 'Obedience' experiment (where volunteers were asked to deliver electric shocks to 'subjects' who got answers wrong, and an alarming number complied, despite the 'screams' they heard from the adjoining room) or David Rosenhan's 1972 'Diagnosis' study, where he asked his students to fake auditory hallucinations to get voluntarily admitted to mental institutions and then see how long it would take for them to be re-diagnosed 'sane' (many weeks in all cases). These studies seemed to shine a new light on the world, compared to even the most acclaimed new theories of economics. Economics – well, classical economics at any rate – has suffered from a crisis of confidence after failing to predict many of the recent financial upheavals. It appears to struggle to make sense of what is clearly not a rational, predictable, 'perfect' world. It is no coincidence that the greatest innovation within this academic sphere, the emergence of behavioural economics, has its roots in the study of Psychology.

This may be because Psychology itself has gone through some significant changes. Academics have devised innovative techniques to measure what was previously thought immeasurable. So, for example, rather than ask people explicitly for their attitudes, we measure their implicit attitudes indirectly via their reaction times to different word associations (a full explanation of Implicit Attitude Testing can be found in '*Blink*'[19]). Economics has begun to embrace the idea that human psychology has a significant influence on how markets operate. Psychologist Daniel Kahneman won the Nobel Prize for economics in 2002, and more recently Stephen Leavitt (who co-wrote '*Freakanomics*') or Richard Thaler and Cass Sunstein (authors of '*Nudge*') have shown how psychological theory and experimental design can offer new insights into the way economic systems – both macro and micro – actually work in real life.

Why we do the things we do

Sigmund Freud studied neuroscience, but became frustrated by the limitations of the physical brain to explain the complexities of the human experience. When he proposed, more than a century ago, that "'*most of our mental life operates unconsciously and that consciousness is merely a property of one part of the mind*' he was vilified by the scientific community. Yet those two hypotheses, that most of our mental functioning happens at an unconscious level and our conscious brain is relatively unimportant in the wider scheme of things, are readily accepted by that same community today.

The traditional view of the human condition is;

THINK → DO → FEEL

We think things through rationally, act on the decisions we take as a result, and then possibly feel the emotions consequent to that decision, following the experience.

Current psychological theory has determined that the more accurate model would be;

FEEL → THINK → DO

Or even more likely;

FEEL → DO →THINK

That is, emotions generally kick in first and create the impetus for our subsequent behaviour, even before the rational mind can respond. The thinking part of our brain will, more often than not, be used to post-rationalise what our emotions had led us to do in the first place, best be summed up by Rory Sutherland, recently President of the IPA, when he said;

> *"The conscious, rational brain isn't the Oval Office. It isn't there making executive decisions in our minds. It is actually the press office, issuing explanations for actions we've already taken."*

Unfortunately, contemporary marketing is still in thrall to this outdated scientific approach. We are too pre-occupied in looking for the 'formula' that will predict and shape consumer behaviour and the 'message' that advertising needs to contain in order to communicate; so much so that we are loathe to admit that human beings don't easily fit into the most sophisticated formulae and the 'Message Myth' is exposed. Our rational minds struggle to understand how feelings can shape the major decisions we make. Recently, pioneering work by research agencies including Brainjuicer, Conquest, OTX and Duckfoot demonstrates the powerful influence emotions play in our consumption patterns and purchase decisions by borrowing techniques from psychology and neuroscience.

By working on our emotions rather than our rational brain, TV influences our behaviours far more than the traditional rational-based theories of marketing could ever explain. How we feel determines what we decide to do; cognitive, attentive thought has very little to do with it.

Decisions, decisions...

We've learned more about how the human brain works in the past decade or so than we had learned in the previous couple of centuries; which is ironic, when you think that one of the biggest breakthroughs in our understanding occurred over 150 years ago, in New England.

In 1846, a railway foreman called Phineas Gage, while laying down new tracks using explosive devices, managed to get a tamping iron, or large iron rod, driven through his cheekbone and out through the top of his skull. He sustained massive brain injuries, particularly to the left frontal lobe, but somehow survived. He spent much of his later life being exhibited as an amazing example of man's ability to survive even the most destructive forces of nature, a man who should have been dead by rights.

He was able to relate the details of the accident straightaway, could almost immediately walk unaided and, after just two months, he was completely pain-free and living independently. The prognosis from his doctor at the time stated Gage *""appears to be in a way of recovering, if he can be controlled."*.

The last point was key; even though Gage appeared to be physically recovered, he was behaving in a strange manner. He'd become impulsive, stubborn and impatient, so much so that, upon recovery, his friends declared he was *'no longer Gage'*! He appeared to have undergone a major change in his ability to feel emotions, manifested in inappropriate or reckless behaviour, because he had lost the ability to feel shame or fear. The extent of his change in personality opened up the possibility that different parts of the brain were responsible for different localised functions, not just motor or cognitive functions, but ones relating to emotion as well. He is certainly one of the first recorded cases where damage to specific parts of the brain resulted in major changes to personality and behaviour.

As well as being difficult to be around, Gage experienced a sudden change in his decision-making abilities; one minute he would be creating grand plans with an air of abandonment, the next he would be vacillating and pulling back, switching his focus on yet another unfeasible scheme. In short, he couldn't make decisions.

In 1994, a young neuroscientist called Antonio DiMasio, used Gage as an example to show a hypothesised link between the frontal lobes, emotion and practical decision-making. Di Masio, in his book *'Descartes Error*[20]*',* demonstrates the relationship, using patients with similar frontal lobe damage. 'Eugene', a previously high-functioning executive with similar brain damage, was recorded attempting to decide on the date for his next appointment. There was no shortage of rational thought in his deliberations; the problem was, there was <u>only</u> rational thought available to him. Without the option to go with his 'gut' feeling, Eugene spent long periods of time weighing up every possible element in his decision, rendering him incapable of making one. This confirmed the importance of emotion in our daily lives; not as a limiting pull on our rational power, but more as a means of sidestepping rational thought altogether, in order to make decisions swiftly and effectively.

DiMasio labelled the emotional short-cuts we use to facilitate decision-making *'heuristics'*. These become increasingly important in a choice-filled world. They enable us to cut out the cognitive effort and make our decision-making manageable.

They are not limited to the trivial decisions, but are often used to help us reach some of the most important decisions we take. Our consumer behaviour is often based on them, but we are rarely aware of them; instead they are fuelled by the brand associations (often called 'brand engrams') and the 'somatic markers', or sensory associations (e.g. between 'Coca Cola' and 'Santa' and 'Red') that Di Masio identified to explain the role emotions play when they move from our memories to our actions.

Through emotion, engagement, memory and heuristics, television can have a huge part to play in those decisions.

It's all about evolution

These new insights help to explain aspects of human behaviour and perception that would not easily fit into the rational model. It helps to explain why it only takes us 17 minutes to decide to buy a house we like; why the smell of baking bread or brewing coffee makes us significantly more likely to buy it; why the colour of a car is more important than the technical specification in our purchase decisions; indeed why all of our decisions, even the most important ones we ever make in our lives, are ruled by emotion.

Let's take our choice of higher education as an example. One would expect that the main influences would be academic reputation and standards, availability of courses, or even more prosaic factors such as distance from the student's home town or local connections. When researchers at the University of Sussex[21] conducted a meta-analysis of all of the available data, it took several sweeps before they finally hit on a key influence that had not initially been considered but which outshone all of the aforementioned rational factors; it was whether or not it was sunny on the day the prospective students had first visited campus. Possibly the most important decision anyone can ever take, and it boils down to the vagaries of the British weather (a shocking basis for any decision!)

It's all to do with evolution. Early man HAD to base his actions on emotions in order to survive; if he encountered something long and thin lying in the grass before him, he wouldn't have taken the rational view, stopping to investigate if it was a harmless stick or a poisonous snake, possibly receiving a fatal bite for his troubles. The basic emotional reaction of fear and aversion would have kicked in before his rational brain had even thought about it and his survival was assured…at least until that sabre-toothed tiger came along!

Neuroscience has since taught us that the emotional brain always responds first to a stimulus, that we can use powerful emotions such as shock, fear, surprise or love to create a behavioural reaction within half a second of the senses receiving the stimulus. The rational brain would usually take up to twice as long, often too late for survival in the primitive world and too late for adequate, timely reactions in today's hyper-stimulated consumer society.

Take a simple purchase - pasta sauce. In an average supermarket, given the range of brands and variants, a fully rational decision on which one to buy could require 36,000 calculations. The weekly shop could take several weeks to conduct at that rate. Our gut feelings and ingrained preferences usually drive the decision (most psychologists believe 90-95% of decisions are made implicitly, emotionally and below the conscious radar). That is where TV really influences purchase, creating those emotional associations that inform and influence our 'gut' decision-making rather than hammering a 'message' into our brains.

There is another way the human brain has evolved to aid our survival; it is all to do with which part of the brain we use and how much it takes out of us.

The 2-stage brain

In one fundamental way, we are all bi-polar; we have not one but two distinct brains, occasionally working together but in a very different way. One of these brains (known as Stage 2 brain) is quite linear, has limited capacity, and often struggles to juggle complex data. Most importantly, from a purely biological point of view, it takes up <u>40%</u> of our body's energy whenever we choose to deploy it. The other (Stage 1) brain is much more flexible, infinitely 'bigger', longer-term and only takes up <u>9%</u> of the body's energy when it is in operation. Guess which one we choose to deploy most often?

It's evolution again. In Neolithic times, it would have been essential that the body used up as little energy as possible. Nowadays, it is the only way we can mentally cope with the cognitive overload of modern life.

The importance of the Stage 1 brain (often labelled the 'implicit mind') should not be underestimated, just because it tends to operate much of the time below our level of conscious thought. It is estimated that the 'explicit' Stage 2 mind (sometimes a filter, but often completely by-passed) is capable of holding approximately seven items of information for a few minutes at a time. Studies on the Implicit mind demonstrate that it holds information many millions of times that capacity; that the information is stored in rich and complex ways; and the 'data' can be held for years, decades even, with little apparent degradation in recall. Experts in this field continue to explore the time and capacity limits of this still relatively uncharted region of the brain and it continues to amaze them just how much we know that falls beyond the range of our conscious awareness.

These insights all help to explain why our Stage 1 brain has become our default position. We are known as 'cognitive misers' by psychologists, as we only put the cognitive brain to work if needed. Meanwhile, the implicit mind does its job in an incredibly effective manner (incredible because we don't even know it's doing it most of the time), helping us to live much of our lives on autopilot. It's the reason why we <u>know</u> to brake a car the moment we see a ball bounce out of a driveway (and well before the explicit mind kicks in) or why we can be in deep conversation with somebody at a dinner party and a barely audible mention of our name from somebody in the opposite corner of the room immediately registers. It's the reason why our instincts about people are much more often right than wrong, despite the inconceivable amount of data we process to reach those conclusions.

Freud was correct when he asserted our unconscious mind was the most powerful influence on the human condition. It affects how we feel, process, experience, decide, remember and behave. Television's influence has been primarily at this level, but it was still invisible to us when the 'TV is dead' narrative took hold, because we weren't looking for it and didn't have the means to measure it. Now it is becoming clearer, as we gain the tools to see the unconscious mind in action.

Market research and the elephant in the room

Market research is fundamentally founded on the Stage 2 brain. Its tradition lies in survey questionnaires, usually based on highly rational assumptions, and focus groups based on...well, focussing - on things such as 'product benefits', 'attention levels' and 'paths to purchase', which almost always force the respondent to think in a Stage 2 way.

In many ways, the online revolution, which has transformed the way we conduct and analyse market research, has skewed things even further - questions are often displayed in very rational, grid-like formats and actively encourage respondents to think about each answer, almost forcing them to deploy their explicit over their implicit minds. Such methodologies are flawed in their ability to predict, diagnose or understand consumer behaviour. They direct the wrong questions to the wrong part of the brain.

Rory Sutherland puts the limitations of market research into context, in a recent article in Research magazine[22];

> *"Behavioural economics shows that affect (emotion) has a strong effect on behaviour, and the mood we're in and all manner of contextual variables affect the decisions we make. Since the majority of money spent on market research is spent on de-contextualised market research – which is essentially asking people to make a theoretical decision – how much predictive value does this really have?*
>
> *We have to ask some questions about how market research can be significantly improved in light of what neuromarketing and behavioural economics teaches us. The value of research doesn't diminish in the slightest – if anything it's heightened – but the nature of the research that we do has to be questioned."*

New research methodologies indicate just how much market research misses out when it focuses on the explicit mind. Implicit Attitude Testing often shows preferences and attitudes that are completely opposite to the 'rational' questionnaire responses. Ethnography demonstrates how much of what people actually do can be forgotten or misremembered when they are asked about it later. Emotional metrics act as a much more powerful means of predicting the future than traditional 'left brain' metrics such as awareness or brand preference. Neuroscience consistently demonstrates that the parts of the brain that are most activated when the foundations of future behaviour are made are the least accessible when people are asked to talk or write about them in subsequent research interviews.

The author of the most ambitious study ever into advertising effectiveness (which I'll cover in depth in the '*Payback*' chapter), stated conclusively;

> *"If there's one measure that seems to be better than any other at predicting the effectiveness of an ad, it's liking of an ad. It's certainly better than communication scores or persuasion measures, and that again links in with this idea that it's emotional engagement, not rational messages, that drive business success"*
>
> Les Binet, MD of DDB Matrix and co-author of 'Marketing in the Era of Accountability'[23]

If market research has consistently undervalued the elements of the Stage 1 mind that are now shown to link most effectively to what products we buy and which brands we switch to and from, then it has consistently undervalued television's role in the process. Even though TV advertising consistently rates well in terms of generating rational persuasion and communication scores, this is not where TV advertising <u>really</u> influences. It is not via active memory, attention or analysis but the holy trinity of emotion, engagement and long-term memory encoding. When the three of them work together, TV advertising works. It's the key to our hearts...and mind(s).

Riding the (brain) waves

Professor Richard Silberstein is a kindly, amiable and softly-spoken Australian. He has worked in the top echelons of clinical neuroscience for almost three decades, publishing over 180 academic papers. He has developed a new form of brain scanning that can monitor activity in surface areas of the brain up to 24 times a second with surprising precision, called Steady State

Topography (SST). As well as expanding the boundaries of our knowledge in crucial areas such as ADHD, schizophrenia, anxiety and short-term memory, he has recently turned his supersized brain to the subject of neuromarketing. He won the Advertising Research Foundation's 'Great Minds' award in 2011.

The marketing and communications industries have been using neuroscience to understand what goes on 'under the bonnet' of our conscious knowledge for almost a decade. In 2003, media agency PHD conducted their pioneering 'neuroplanning' study, using a technique called 'functional magnetic resource imaging (fMRI) to look at brain activity during different media experiences. It showed conclusively that TV and cinema content activated two specific parts of the brain; the hypothalamus and the amygdalla, responsible for our long-term memory processing and our emotions respectively. Emotions and memory - where brands live.

Although fMRI is fantastic for measuring deep brain processes very precisely, it suffers from two limitations; it can't measure changes across short time intervals very accurately, because it measures blood flow and there is a delay effect. Also, the scanner used in fMRI is cumbersome, noisy and restricts movement, so it is not very representative of how we normally experience different media. That is where Professor Silberstein's technology comes in.

SST can be measured using electrodes fitted to a headcap. Subjects can walk around whilst wearing them and consume media in a much more realistic way. It measures five key elements of brain activity;

- Attention
- Emotional Intensity
- Emotional Direction (i.e. approach/withdraw)
- Engagement
- Long-term Memory Encoding

In 2010, I commissioned a study while at Thinkbox, using both SST and *f*MRI, to look at what really goes on in the brain when people watch TV and when they are online. We recruited 17 advertisers and put their TV ads, and online content, under the scanner (and electrodes). The results were fascinating.

The main measure, relating to advertising effectiveness is long term memory encoding (LTME), particularly levels of LTME during 'key branding moments', when the brand is displayed or integrated into the content. Previous studies carried out by Professor Silberstein and his team show that this is a major influence on future brand purchase behaviour[24].

In the Thinkbox study, levels of LTME differed significantly between the 17 commercials we tested. Some managed to generate higher levels of LTME when the brand was present, whereas others missed out because the brand featured at the wrong time (the study also gave us some fascinating insights into how TV ads can creatively address this challenge). So, why does the brain decide to process some of the content it is experiencing into LTME, but not others? The answer can be found in the relationship between the four other measures, and LTME itself.

The power of emotion to drive memory is indisputable; the level of emotion we feel when we are watching an ad is strongly related to levels of LTME - a correlation of close to 60% in this study. This strongly supports Les Binet's assertion that liking of an ad has most influence on whether or not the ad works. It also helps explain why television, what Binet referred to as *"the emotional medium par excellence",* comes out so well in the vast majority of effectiveness studies.

Emotional intensity and engagement were also closely related, but engagement had the strongest relationship with LTME, with a correlation of over 70%. This is one of the few studies to really demonstrate how engagement works. It is about people being so immersed in the content that they are processing huge amounts of 'information' (mainly via their implicit, associative mind) about brands, much of which they think they know 'for themselves'. This forms the basis for 'brand memory' which itself has a strong correlation with future brand loyalty and switching behaviour[25].

From neural pathways to pathways to purchase

The associations we have with brands are known as 'brand engrams'. They can be built over a long period of time (think of things like 'Guinness – worth the wait' or Audi – German efficiency – 'Vorsprung Durch Technik'). Once established, it is very hard to replace them (as Masterfoods discovered when they tried to move away from the 'work, rest and play' association for Mars Bars). These associations can be moulded through marketing but their durability is for a simple reason; they tell each of us our own personal *story* about the brand.

However, television isn't just about forming brand memory; it is increasingly about generating action. Over the past few years, TV has seen its role as a response medium explode. We'll explore the reasons for this in later chapters, but there is no doubt that people are responding to what they see on television – programmes and commercials – in unprecedented numbers. People are sending off for further details, voting on talent show competitions, finding information and purchasing products straight from the screen.

In pre-internet days, the time period between seeing an advertisement and purchasing the product could be across days, weeks or even months. The consumer would have to physically visit the store to buy, and the connection between exposure and purchase was somewhat vague. Nowadays, the impact of a TV ad can be immediate and we have the tracking tools to measure it in minutes rather than across days.

At some point, the engagement that an ad elicits will help to generate that response, at which point attention levels do rise.

That is when other media channels, especially online, can harvest the interest and desire into valuable interactions and, ultimately, purchase. It appears consumers need some level of engagement to drive that process, otherwise they remain happily in their low attention state, living up to their reputation as cognitive misers.

We're still at the edges of working out how it all fits together, and I'm currently working with Neuro-insight on a study to plot this route from engagement to purchase, but there is no doubt that, while engagement and attention may not be related, they are not strangers!

Engagement appears to act as a precursor of attention, especially if that results in a transaction of some kind. That heady mix of engagement leading to attention is the fuel behind the rapidly increasing responsiveness to TV content which we're seeing via Google, Facebook, online retail and advertisers' own websites. But without the engagement that TV provides, attention is very hard to activate (as can be shown in the declining response rates for online display advertising).

Why all this matters to television

The way the human brain works isn't driven by logic or rationality, but by emotion, engagement and long-term, associative memory. It doesn't behave in a linear, predictable way but through complex, context-sensitive assimilation. Everything we have learned about how the human brain works in the past twenty years or so has reinforced our understanding of why television is so resilient, influential and popular.

The digital revolution has enabled television to create stronger emotions, greater engagement, more immersive storytelling and deeper sharing of the experience. It has cemented television's role in modern culture, not by changing the way our brains work (many of the assertions that human brains are altering fundamentally as a result of the impact of digital technology are not supported by the evidence) but by appealing to the mental processes that have been driving our thoughts, feelings and behaviours for millennia. The science has finally shown us how it's done; market research is still playing catch-up.

Television works because it is perfect for the way our brains work when we are at our most receptive. Our brains love TV because it is rich, deep, immersive and endlessly entertaining. That is down to the second element of engagement - the content.

CONTENT

The power of audio-visual

In three decades of working in television, I have rarely heard people talk about its main advantage as an entertainment or advertising channel - the depth, richness and communications power of audio-visual. Even though 'rich media' has been heralded as one of the key developments in the online world, its role in TV's endurance in the face of so much competition often gets overlooked.

The way television delivers audio-visual entertainment is going through its own revolution. Greater theatricality, through flat-screen TV sets and high definition broadcasting has transformed the TV experience and the trend is set to continue, with Ultra HD, 3D and improving set technology. Theatricality is the element of the television experience that people value the most. It should also be what advertisers value most.

A recent paper[26] on the power of audio visual to both communicate and generate emotion pointed to the richness of gesture plus dialogue working together. Audio + visual creates far more than the simple sum of its parts. Dr Geoffrey Beattie,

Professor of Psychology at Manchester University, demonstrated how AV was significantly more effective, and longer-lasting, than audio or text only, for communicating information across a range of products.

A lot of this is down to the power of AV to demonstrate product benefits, usage and information; that is the practical advantage of AV. The emotional element can be explained in large part by the theories of meta-communication put forward by Paul Watzlawick– we take out enormous amounts of implicit information and meaning from our interactions with and observations of people, even from the slightest gestures or shifts in tone of voice. Screenwriters such as William Goldman[27] understand this, when they talk about the freedom AV gives them to convey a great deal of character information through a simple gesture, rather than through pages of exposition. Audio visual enables us to instinctively understand meaning, motivation and emotion in its depiction of both everyday and extraordinary human situations. It also enables us to feel what they feel. That is why our emotional brain works on overtime when engaged with great TV content.

AV provides these communications advantages in three distinct ways;

1. The importance of empathy
2. The power of storytelling
3. Rich experiences

1. The importance of empathy

There is something fundamental to human nature that helps to explain television's continued success as a communications medium; the power of audio visual to generate the feelings associated with other peoples' experiences. It is a building block of empathy and it can be understood through a few neurons that appear to fire in conjunction with our observations of other human beings' behaviour, 'mirroring' that behaviour as if the observer were engaged in it themselves. These simple brain processes can have a strong effect on our own emotions and behaviour, causing us to reflect or 'mirror' those of others.

Mirror neurons were only discovered in 1992 and we are still learning about their role in human cognitive and emotional development. Leading neuroscientists believe mirror neurons are vital to the learning process, especially in areas such as language, socialisation and motor development, and that they may be strongly related to disorders such as autism. We can see or feel them at work every day, for example when we yawn just after somebody else has done so, laugh along with the audience (or join in an uncontrollable fit of the giggles) or automatically wince when somebody else feels pain. Recent studies have shown that mirror neurons are more numerous and dispersed than previously thought and their presence has been measured in babies less than twelve months old.

Our ability to learn from others and to feel what they feel is a basic human survival skill and a major influence on our cognitive and emotional development. Audio visual media are particularly effective, for obvious reasons. Although sound and images alone are known to fire off mirror neurons, it is when all the senses are engaged at once that the activity is maximised. That is why, after a particularly emotional or bruising movie, people often leave the cinema with tears in their eyes or a pained shuffle in their walk; what we see on screen deeply affects how we feel. Television's role as a shared medium means it can work on our mirror neurons in two ways; through the emotions and experiences of the characters on screen as well as the reaction of the other people watching with us. It is a powerful combination.

It is not just audio-visual in itself that provides that emotional impact. It is when AV content is structured in a way which is known to massively affect how we feel and what we do. It is most effective when it is used to fulfil a fundamental needs state; our need for a good story.

2. Tell me a story...

It was a speech by John Hendricks, the founder of Discovery Channels, at the 1996 Edinburgh TV Festival, that made me realise how deeply storytelling is ingrained into television's genes. I expected him to talk about the educational and factual quality of his channels, but instead he delivered a heartfelt eulogy to their narrative qualities, how they carried on the campfire storytelling traditions to create a sense of intrigue and wonder about the world around us. It was then that I started to grasp how influential storytelling can be.

I wanted to learn more about storytelling's influence, because that is what television does best; it tells stories, often intertwining, coming at the viewer from all angles. Beautifully crafted 30 second stories about brands mixed up with a complex blend of programme-led stories, often unfolding across weeks and even months. The TV experience is all about story.

Storytelling is the fundamental driver of drama, news and sports coverage, reality shows, great situation comedy and documentary programmes alike. The viewing figures, revenues, effectiveness studies and sheer *currency* television still occupies in modern life all reflect its power in keeping audiences engaged and advertisers are offered a unique environment to extend that engagement through their own brand stories.

Relatively few experts in the media and marketing fields have linked TV's storytelling power to its continued success in audience or revenue terms. As with emotion, it may seem too 'fluffy' or unscientific to be embraced by the marketing sciences. But stories are TV's lifeblood, whether it is the 'journey' that every '*X Factor*' contestant shares with the audience to that same audience speculating why Cadbury's would have the crazy idea of having a gorilla playing drums to a classic Phil Collins tune, and what does it 'say' about Cadbury's? In the latter case, a great deal of the storytelling will go on inside the consumer's head.

I wrote an article for a major advertising journal recently[28] about storytelling's power to influence us, but I was disappointed with how little of my evidence came out of marketing or media. The overwhelming evidence to support the role of story as a fundamental driver of human perception and behaviour came instead from diverse academic fields such as education, social anthropology, psychology and even theology (after all, most major religions are story-based; think of the power of the parables to influence Christian society's behaviour for two thousand years). Studies show that we are born with an innate attraction to story. It profoundly affects our development, education, behaviour, morality and social interactions throughout our lives. Babies under the age of 12 months have been shown to have a grasp of narrative before they utter their first words or walk their first steps.

The importance of story for brands has only begun to be fully recognised by marketers in the last decade or so. It is based around the concept of 'transmedia storytelling'. I first heard the expression from Jonathan Mildenhall of Coca Cola at the 2008 Media 360 event, citing Coke's own 'Happiness Factory' campaign along with others as offering something that is *"much more brand-centric and designed to build 'brand mythology"*. Transmedia storytelling should include a combination of *"an immersive world, heroes, villains, friends and foes, a compelling story and multiple access points"*[29].

This was ironic, because for a while there had been a definite emphasis on the multiple access points and rather less on the storytelling. One sign of the shift in attitudes is that major multinationals such as Coca Cola, Burger King, Ford, P&G

and Unilever now employ storytelling specialists within their marketing departments.

The true power of storytelling can be linked back to an ancient Indian proverb;

"If you tell me a fact, I will learn. If you tell me a truth, I will believe. But tell me a story and it will live in my heart forever"

It is those last two words – *'heart'* and *'forever'* - that provide the key to storytelling's increasing influence in brand communication. As numerous studies in neuroscience, psychology and advertising effectiveness have shown, what goes into the heart (our emotions) can live 'forever', through long-term memory encoding; not as facts or messages so much as powerful associations.

It is rare to see a concept take shape when looking at raw research results, but analysis of the brain patterns in Thinkbox's 'Brainwaves' research clearly depicted the brand story taking shape, in the peaks and troughs of engagement and long-term memory encoding; the better the story, the higher the peaks. Neuroscience graphically illustrated storytelling's link to emotion, engagement and long-term memory in a way traditional research techniques could never accomplish.

3. Rich media = rich experiences

The coalescing of images, colour, movement and sound to create a richer storytelling environment has been critical to TV's hold on its audience since its genesis, but we are only just creating the research and analysis tools to value that richness. Meanwhile, TV's storytelling powers are transforming; the way television content is produced and presented to us has changed more than for any other media experience.

I often illustrate this by showing a short collage of TV programme clips and advertisements, first from the 1960s and 1970s before cutting to the present day. The difference is astounding. TV has evolved from monochrome to colour, from standard definition to HD and from analogue to digital, revolutionising the delivery of TV content. It is bigger, brighter, sharper and more intense.

Meanwhile, advances in digital technology have transformed production capabilities, with more portable and higher quality cameras and infinitely more versatile editing processes, even for the most basic production budgets. This is why even the niche channels can now produce their own content rather than rely on imports and archive material. It is also why big budget TV series have become almost cinematic, with often astounding production values, prompting many major film directors to express a preference for working in television.

Aesthetically, the content just gets better and better.

TV is also becoming ubiquitous as audio visual becomes the communications medium of choice. In the pre-digital age, the cost of a TV campaign could only be amortised across TV advertising. Now, the same audio visual (or a version of it) can be shown on mobile phones, digital outdoor, laptops, tablets and at digital point-of-sale. We're becoming an audio-visual world, and TV advertising is no longer just for the TV screen.

Why, then, haven't other forms of rich display media (especially online display) started to seriously compete with TV advertising? It may be to do with a number of things, including scale of audience and quality of content. Even when online can replicate the content, however, it cannot always replicate the context; we still prefer to consume this form of entertainment on a big screen in a comfortable location, based on TV schedules that are curated to allow us to lie back and be entertained. TV becomes more powerful because of the context of viewing – we are in a relaxed, receptive mindset and consuming the content within a familiar and comfortable physical environment, shared with the people we love. As well as content, the <u>context</u> of viewing is influential, more than we ever previously considered.

CONTEXT

"For humans, everything is relative, there are no absolute measures...our judgement becomes swamped by context"
Professor Nick Chater (UCL)

The third element of the ARF's definition of engagement is context, the power of the surrounding environment. It has been the hidden persuader for many years, but marketing is finally waking up to its importance.

The Thinkbox Engagement Study, which I mentioned in a previous chapter, was the first rigorous analysis of what really happens when people are watching TV, especially around the ad breaks. It was not just the qualitative value of having thousands of filmed observations to individually assess; each observation was coded against 117 variables relating to audience, content and context of viewing. This produced a quantitative analysis based around almost 2 million data points, an astounding sum for an observational study.

The Engagement Study produced a raft of insights that could never have been achieved by self-reporting; in fact, many of the deeply engaged behaviours that the participants exhibited were conducted well below their levels of awareness. The study demonstrated that, when people were primarily viewing TV during the ad breaks, engagement-related behaviours (eyes switching to screen, people talking about the ads, non-verbal gestures etc.) could be identified in more than two thirds of occasions, and a clear majority of such behaviours were positive (43% compared to 26% negative). It demonstrated that engagement can be seen in a range of behaviours. And this, remember is only what could be observed.

Perhaps the most surprising finding from the study was the power of the context of viewing. There were two elements to the power of context in influencing engagement; the surrounding on-screen environment and the surrounding physical environment. What, when, where, how and with whom people viewed television had a significant impact on how they engaged during ad breaks.

On-screen context

The surrounding on-screen environment's influence on audience mindset and receptivity has been much explored. Advertising research, for example, often finds a halo effect on advertising effectiveness from the 'quality' of the surrounding programmes[30], but the measures of effectiveness most often used (e.g. day after recall) are far from the best indicator of the way this power works, and such studies tend to show a fairly weak or inconsistent relationship.

The Engagement Study suggested a much stronger on-screen influence, driven by the viewer's desire for congruence and quality, from not just the surrounding programme content but also the surrounding advertising. Positive engagement behaviour increased during programmes that people also found engaging (a phenomenon also seen in the 'Brainwaves' study), as evidenced by the significantly higher engagement levels during peak-time programmes. Engagement around TV ads also tended to happen in clusters, so engagement elicited by one ad could be maintained into subsequent ads in the same break. Ad and break positioning were also important, and ads that feel congruent or in tune with the surrounding programme content also tended to increase engagement, a finding that recent studies from Viacom[31] and GMTV[32] have reinforced.

Off-screen Context

The real-life, physical context of viewing also influences engagement, with variables such as daypart, number of people in the room, location, concurrent activities and technology used (e.g. viewed live or timeshift) showing a significant relationship. Engagement is most likely to peak when people are watching on the main TV set, during peak time, watching a programme with which they are highly engaged and in the company of other people. The number of people in the room is a particularly interesting part of the context. Two people in the room provided the optimum number; they would often direct each other's attention to what was being advertised. Upwards of three people, however, the average amount of engagement observed declined, as social interaction became more of a distraction. Even then, advertising was often able to cut through into the conversation - when it did, it could set the whole room talking.

The study was the first time we had on record just how much 'chat' occurs around television, both programming and advertising. It demonstrated the power of the shared experience that is still unique to television and is now being enhanced by technology. We'll explore this in more depth in the next chapter ('Social').

Conversations about one brand make the point succinctly. A well-known breakfast cereal ad, featuring a grating sing-along jingle and overly cheery kids, led to conversations in two different households that, in total, featured the brand name seven times, the manufacturer twice, discussion about the lead actor and a whole conversation about whether the brand was better served with hot or cold milk on winter mornings.

The power of peak to prime the audience

It was not surprising that most positive engagement behaviours occurred during the daypart when conversations take place in large numbers, when consumers are relaxed and receptive enough to chat, to reflect, and to engage, at the end of the working day. That time is the most precious jewel in television's crown. That time is peak time.

Peak-time (or prime-time) has always been able to command an advertising premium; hence its name. Peak-time's ability to reach the bulk of the population quickly cemented its status during the first age of television. Its ability to create a special, more receptive environment for advertising to work effectively sustains its value during the third age of television. But, no doubt about it, peak-time is special time as far as influencing consumers is concerned; indeed, peak-time is television time.

In 2007, the UK's Institute of Practitioners in Advertising (IPA) embarked on a research programme to measure how people use media in the digital era. Their initiative, Touchpoints, is a remarkable insight into our increasingly meshed media world. For each survey, over five thousand people are asked to record what they are doing each half hour of the day across two weeks. It covers their media use and most other main activities in fine detail. Its main role is to provide a 'hub' survey, connecting all of the industry media research vehicles to allow integrated media planning, but it also allows us a fascinating insight into people's daily lives and the increasing role media play.

Television has the biggest role of all, at least if we use time as our barometer. The 2010 Touchpoints survey gave TV a 50% share of our total media time, with online time per person at 19% (it was 13% in the original 2007 survey). Across those three years, reported hours of TV viewing have remained rock solid, reflecting the BARB data. What has happened is that more of our TV viewing time is shared with other media – especially online. As we shall see in later chapters, this is a major opportunity for both programme makers and advertisers.

Figure 1

Share of media time by media channel – All day vs. Peak-time

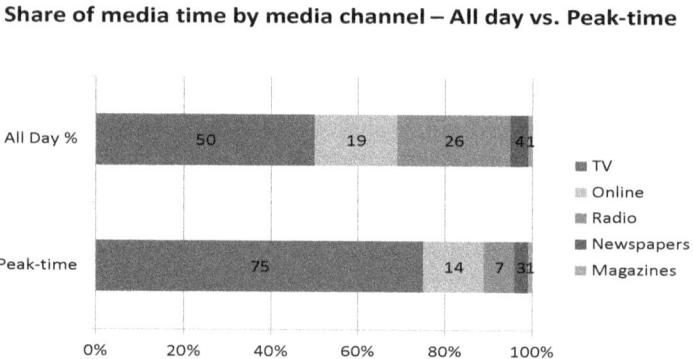

Source: IPA Touchpoints 2010 – Adults recorded media time (Peak-time = 7pm – 11pm)

Reproduced with kind permission of the Institute of Practitioners in Advertising

TV's dominance of our media day is driven by peak-time. It is the time when people gather, relax, switch from work mode to play mode and adopt an 'entertain me' mindset. TV's 50% all-day share translates into a 75% peak-time share (i.e. 7pm to 11pm). TV not only dominates media usage during that special time, but it also drives a great deal of online and other activity.

We need to think about that for a moment, because we've

tended to see 'peak' as just another daypart. It is a time when many millions of people will be watching TV, including many who cannot be reached at other times of the day. They are in a more engagement-ready mindset and they are watching higher quality content. They are more likely to be sharing that experience, either in real-life or virtually. It is a prime space for advertisers to inhabit because it offers something unique, which is why they have been willing to pay a premium to advertise there and why broadcasters have always spent a disproportionate chunk of the programming budget on it; even before either party had the tools to understand just how much that premium was worth.

SOCIAL

Take a look at any image of typical TV viewing from the first age of television and you will see the whole family huddled around the glowing box in the corner of the room. They were huddled because the screen was small, living space was often cramped and – in the early days at least – central heating was still a pipe dream. But there was always a family group because TV had been designed to create a shared, communal experience.

Until recently, the shared viewing experience had been consigned to the scrap-heap, a natural by-product of fragmentation. Owning a second TV set had been a rarity among most households until the 1990s, whereas now television content is available on a variety of screens, spread like scatter cushions throughout the home, with dozens of channels to suit each individual's taste. So, it would not be surprising if shared TV viewing had dropped exponentially over the past couple of decades and almost disappeared by now. In fact, the very opposite seems to be happening.

The Engagement Study recorded three types of TV viewing – **in-between time** (when TV would be an accompaniment to other activities, such as housework or studying), **me time** (watching alone, able to engage in personal viewing choices), and **we-**

time (where we watch together, compromising on our choice of programmes). Despite the predictions that we would mainly be viewing personal content alone by now, technology has barely made a dent on how much TV we watch with others, nor on its ability to get us talking. We-time TV viewing is the predominant way we view TV, because it is the most engaging.

A recent analysis of BARB viewing data[33] from 2005 to 2010, a period when TV viewing patterns were supposed to have changed dramatically, shows a surprising consistency. In spite of the proliferation of viewing choices, the proportion of TV viewing that was shared with other people in the room had remained rock solid (i.e. around 70% of all individuals' viewing) across the five years. The consistency was surprising in itself, but when we looked at the changes in household composition across that time period, it was truly staggering. Shared viewing stayed steady despite an increase of 16% in the number of single person households. So, the amount of shared viewing must be rising significantly in multi-person households OR single person households are finding ways to share their viewing experiences with others.

That is yet another example of how the prevailing narrative has been proved wrong. Personalisation was supposed to trump everything, boosted by the availability of screens content and interactivity, so why would we continue to choose to watch with others in shared spaces, with all of the negotiation that entails?

This obsession with personalisation comes from the technology sector, not from real people; maybe technology creates a solitary mentality and technologists tend to live alone. Where TV is concerned, personalisation is far less of a priority than compromise. Finding sharable content is often deemed much more important.

Personalisation is technology-driven. Although it is becoming possible to watch TV anywhere, on any screen, we still often <u>choose</u> to watch on the main TV set. This is mainly because it is the best screen available, and all of the channels and content can be found there (e.g. most pay TV homes only have access to premium content via the main set).

Consumers also design their lives around the television. Just last year, German TV sales house IP Deutschland studied the design of more than 1,000 living rooms in Germany and observed that all but a handful had been designed around the TV set[34].

As we learn more about the inherent need people have to 'be part of the herd', it becomes increasingly obvious why TV should be integral to the shared experience. People need something to talk about, and experience together. That, in turn, guarantees the social fuel for their conversations. This is becoming increasingly true of programmes and commercials alike.

TV's role in our daily lives

Despite the regular outbursts of anguish about the breakdown of modern family life – with television often cast in the role of chief villain – the reality is that families spend a great deal of time together. Many of the rituals that are supposed to be dying are still very much a part of that family life and television still appears to be the default option when members of the household wish to 'hang out' together.

If we look at that valuable time after the working day, when household members often congregate, then (according to Touchpoints) we still regularly eat meals together at the same time (well over half of all families eat a meal together on weekday nights); we watch TV together in large numbers from just after the evening meal until bedtime; and we also engage in many other behaviours together – such as conversation (face-to-face, by telephone or online) – which are often about the programmes being broadcast or the brands being advertised.

I mentioned earlier that online currently accounts for just

under a fifth (19%) of our total media time. During peak-time, online takes a lower share – just 14% – owing to television's dominance, but more than 90% of that online time is spent in front of the TV set, the two being consumed either concurrently or sequentially. This is a major opportunity for both broadcasters and advertisers alike, as we shall see. When I first demonstrated the scale of this phenomenon in 2007, Sue Unerman, Chief Strategy Officer of Mediacom, responded by stating *"this makes television a point of sale medium!"*

One of the more established rituals nowadays is still around 'appointment-to-view' television, a phrase that was coined when people had to organise their lives in order to watch a particular programme, as there were no opportunities for catch-up, either via on demand or through timeshift. Until the last decade or so, many viewers would plan their week's viewing well in advance and often organise their personal and household schedules around key programmes they wanted to watch together.

The whole concept of appointment to view was supposed to have faded away by now, but as we have already seen, the live schedules are being watched by record numbers of people and even a great deal of non-linear viewing is believed to be appointment to view because people are using new technology to make appointments to view together. Such is the power and allure of the shared viewing experience.

Another form of shared viewing that is part of people's daily ritual is virtual shared viewing via the telephone or online social networks. There is significant anecdotal evidence that many people are arranging to watch a programme live – or synchronise their timeshift viewing of a favourite show – so that they can enjoy it with friends and family not in the room with them. This is particularly true of event programming, but also extends to soaps, reality shows (the *'Next Top Model'* series appears to receive a disproportionate number of mentions in my research) and major dramas.

All of these examples demonstrate the need people have for shared experience, ritual and structure, and how integral TV is (and has always been) to those needs. As with so many other examples in this book, it had been widely assumed until recently that these behaviours would have been severely disrupted by the new technologies that promote personalisation and offer so many different alternatives to viewing together, and yet all of the evidence is to the contrary. Shared television viewing has become ingrained, established and enhanced by the very technologies that were meant to kill it off.

Broadcasters are beginning to respond to this opportunity. TV schedules already look very different to the pre-digital era and there has been a major shift towards lifestyle, reality, sport and 'event' programming. Programmes in these genres succeed in driving viewers to the live schedules and getting them talking about the content. In short, they are not just appointment to view, but also appointment to <u>share</u>.

Reality TV is a case in point. Initially, it was seen as a low cost alternative to more established genres, as digital technology made production cheaper and more flexible. Digital filming is able to 'capture the moment' in ways that the cumbersome analogue cameras could never manage. Reality programming often has to be seen live to be fully enjoyed (just like sport). It can be spun out into many different formats, can contain lots of side programmes (*'Big Brother's Big Mouth', 'The Xtra Factor'*) and, perhaps most importantly of all in the current climate, it gets people talking.

The talkability factor is a relatively new one. We have known for decades that great television (or even really bad television) will get people talking about it. We intuitively understand the phrase 'water cooler television'...and we don't even have water coolers in the UK. We see the conversation reflected in the letters pages of the national newspapers, on online forums, via magazine articles, or even via our own conversations. It took the emergence of the internet, though, to provide a glimpse of the widespread, viral and immediate nature of TV-generated word of mouth.

We dismissed talkability as an outcome rather than a driver of human behaviour, if we thought about it at all. We didn't understand the value of social currency. Now we do.

Word of mouth – the new television?

In 2009, I attended several industry conferences about the future of advertising. As social media had really begun to take off and advertisers were beginning to adjust their budgets accordingly, one of the gurus of social behaviour ended his presentation each time with a chart headed '*Word of mouth is the new television*'

I find it hard to know where to begin with that statement, apart from making the obvious point that something that is so dependent on something else, cannot then be considered its replacement. In other words, word of mouth is <u>not</u> the new television, because without television, word of mouth would be severely diminished - especially word of mouth pertaining to brands.

We have always talked about what we see on television. Big events, major scandals, celebrity misdemeanours, soap tragedies, hard-hitting dramas and cutting-edge comedy have always been among our favourite and most ubiquitous topics of conversation, taking place on the morning commute to work, during coffee breaks, in the pub, at school and all the way to bed. It comes naturally to us because television is one of the few topics that everybody has experienced and on which almost everybody has an opinion.

Whatever term we use - word of mouth, chatter, buzz, conversation, recommendation - it is a key part of being alive and being connected. Sometimes, though, we disappoint the talk experts. It is expected that we should use our time productively, to discuss matters of importance and depth, but we rarely leave the shallows. As we saw in the '*Audience*' chapter, our default mode is 'easy', working on auto-pilot, unwilling to engage the energy-sapping, cognitive brain. Our conversations follow the same pattern. Even when people may wish to self-represent themselves as more serious, deep individuals, they still cannot lie when it comes to what they talk about. As the long-running Television Opinion Monitor (TOM) survey[35] reported, they talk about each other first and television second, and that has been the case for at least 30 years. The world around us, the economy, the political climate, the environment - such serious topics tend to get relegated to the back of the bus.

The Thinkbox Engagement Study, with its thousands of filmed records of people watching commercials in their natural habitat, showed how conversation happens *in real life,* rather than how people report it after the event. It showed conversation – about people, brands, content, life's distractions and amusements – comes from all directions, has a wonderful random quality and cannot be predicted or controlled. In this supposed age of viral and social behaviour, it became apparent that there is nothing more viral or more social than two or more people sitting together, talking about the experience they are sharing on-screen. The problem has been that, until now, it has been generally invisible and therefore forgotten.

One of the best clips from the Engagement Study has a young couple, soon to be married, watching TV together late one evening. The conversation turns to the news that two of their best friends are about to have a baby, which seems to be leading to one of the most important conversations any young couple is likely to have - is it time for us to be thinking along those lines too? As the discussion develops, one of the most creative ads of that year – the Guinness Cold '*Evolution*' ad – appears and they immediately switched conversation to discuss how much they loved the ad. There follows a wonderful exchange of dialogue where the entertainment value of the ad, the Guinness brand itself and their future family planning priorities became interweaved.

These conversations have a huge effect on how we perceive brands and remember advertising, but we are rarely asked about them in market research and, if we were, it is highly unlikely we would remember them. Fortunately, as the word of mouth phenomenon has finally been recognised as an important element of marketing communications, we now have a range of vehicles to measure it.

Ed Keller is a highly respected marketing expert who has been called *"one of the most recognized names in word of mouth"* following publication of his book *'The Influentials'* in 2004. Together with his partner Brad Fay, who has an equally blue chip reputation within the market research industry, they founded Keller Fay in 2006 and immediately set up TalkTrack®, a continuous measurement vehicle for word of mouth about brands. They interview 36,000 Americans every year (and now a similar number in Australia and the UK) providing data on over a million individual brand conversations every year, including information on where the conversation took place, which markets and brands were discussed, and whether it was a positive or negative mention. As the respondents log their conversations as they occur, it is far less subject to problems of reporting error or forgetfulness.

TalkTrack shows that word of mouth does not work in the way we have been led to believe. For example, out of the millions of brand conversations recorded, the proportion generated online is just 7%, a figure that has been consistent since TalkTrack launched and replicated in the UK and Australian studies. The vast majority of brand conversations occur through face-to-face conversations, although telephone accounts for around twice online's contribution. In other words, as with many phenomena reported in this book, online has not replaced anything, but it has turbo-charged human behaviours that have been going strong since well before the digital age.

The study also provides an insight into why word of mouth is now considered so important. It is prevalent - people report on average 60 brand mentions a week. It is generally positive – people talk about positive aspects of brands around three times as often as they talk about the negative. It is diverse and wide-ranging - we talk about day to day brands such as Coca Cola, Starbucks or major retailers rather more than the digital stars such as Apple or Google, with food & drink, health & beauty, technology and financial services being the most popular categories. Perhaps most comforting of all to advertisers, it proves that marketing works - almost half of all brand conversations are prompted by marketing, with television (17%) and internet (15%) the most cited sources by far.

Keller and Fay have expanded their analysis in their recently published book *'The Face-to-Face Book; Why Real Relationships Rule in the Digital Marketplace*[36]which is full to the brim with examples of how it is our relationships in the physical world that power our decisions, influence our thinking (and, most of all, feelings) and determine our brand preferences.

Our conversations define us and define our relationships. Brands have always featured deeply in those conversations and marketing has often been the spark. The idea that the power of word of mouth has only really been unleashed with the advent of Web 2.0 is, frankly, ludicrous, although the transparency and accessibility of online data has brought it out of the shadows. Now, we need to understand how to best manage and integrate it, because it can be directed (to an extent) and it does work around our media experiences, rather than as a replacement for them. Innovations like TalkTrack have demonstrated not that word of mouth is the new television, it's not the *new* anything, but that television is one of the main generators of word of mouth and online is one of many ways it can be harvested. Things have always been that way, because we've always needed things to talk about. If the conversation is seen as a <u>replacement</u> for television, things can go very badly wrong indeed...

The Pepsi challenge – creating word of mouth without TV

Pepsi had been calling for ideas to support its push into social media as far back as 2008. This resulted in the Pepsi Refresh Project, launched in early 2010, funding small community projects based on online votes. The project was funded primarily – and very publically – with money taken from Pepsi's TV budget. TV had been sacrificed for social media.

By spring 2011, PepsiCo's CEO reported on progress at TED. The RefreshEverything.com site attracted more unique users on a monthly basis than a host of other sites where the brand would have previously advertised. In addition, the project attracted 7,500 applications in its first year, and 80 million votes, the brand's "likes" on Facebook had increased tenfold to 3 million and its Twitter following (600,000) and YouTube presence had also grown. These were impressive online numbers – and, furthermore, Pepsi has been praised for the social good that many of these activities create.

However, not everybody was happy – especially Pepsi's shareholders. They had seen their main brand drop to an unprecedented third place in cola sales, falling behind Diet Coke for the first time in history, with an overall sales slump worth up to half a billion dollars. This has been commonly attributed to the lack of branding or even visibility in the market, allowing Coca Cola to steal market share at an unprecedented rate.

Here was a genuine Pepsi challenge.

This is a great example of the binary thinking that still

too often exists; it's one thing *or* another. Marketers need to concentrate on *and*, not *or*. If we look at the winners in the cola wars, we can see what happens when binary thinking makes way for a more nuanced and integrationist approach, as was the case with deadly rivals Coca Cola.

It wasn't as if Coca Cola merely responded by betting the house on TV, indeed it had been rightly praised for its social media presence even while Pepsi was floundering. But, Coca Cola integrated its social activity in line with its TV activity. A good example was its 'social zone' in support of its TV sponsorship of the NCAA tournament. The result has been a highly visible presence, a well-understood brand position (The Happiness Factory) and a boost in market share.

It has certainly caused a shake-up in Pepsi's thinking. In November 2011, Shiv Singh, the Global Head of Digital for Pepsico Beverages, defended the power of the TV spot advertising, arguing that it should be seen as a trailer for a deeper digital experience, insisting that "*in the future, TV spots will be most effective when woven into the fabric of digital media*". More recently, at the 2012 Festival of Media, Salman Amir (Pepsico's executive VP and CMO) said;

> "*So, what is that next big thing? Is it something to do with social media? Or mobile? Or is it some new advanced technology that somehow I, and I alone, have been given an exclusive sneak preview into? For me, the answer is none of the above. It's just television*[37]"

Of course, this does not spell the end of 'social media' as a marketing channel – that would be a ridiculous conclusion; but it should hopefully hasten the end of the ridiculous binary approach. We need to focus on how social media can be best integrated with other marketing channels, rather than seeing it as an automatic alternative to them. After all, conversations don't happen in a vacuum but the kind of binary thinking can create a vacuum in terms of brand visibility and emotional engagement.

Why do we need social fuel anyway?

Our need for conversation is as fundamental as our need for narrative structure. We are born with it and use it as naturally as breathing.

Conversation – social fuel – has a number of functions. We may have conversations based on rational information; we need to organise our lives, ask for directions, request information or compare prices. However, a surprisingly large percentage of our conversations concerns the more trivial or emotional elements of our lives. We like to gossip about family and friends, share our feelings, speculate on the lives of celebrities, discuss the weekend's sporting headlines, tell jokes and engage in idle banter, far more than we would like to admit. Such conversations rarely emerge from market research; it's not how we like to represent ourselves and, besides, market researchers rarely ask questions about them!

Research I conducted for a major newspaper group showed quite clearly that readers were more influenced by the social fuel their newspaper could provide than by the political slant of that newspaper. This helps to explain why so many Labour voters read The Sun (and quite a few Conservatives read the Daily Mirror, or Liberal Democrats read The Daily Mail!).Those newspapers are often valued more for the items that spark comment than for their insightful political analysis.

Social fuel is an important part of most people's lives for several reasons, and television is capable of addressing them all.

It is important because we are inherently social animals, and we need social fuel to maintain and enhance our relationships. In his book *'Herd'*[38], Mark Earls goes even further, arguing that the 'i' emphasis in Western culture is something of an aberration and we are programmed as a species to be more in tune with the 'we' philosophy of many Eastern cultures. Earls points out that our social interactions nurture, educate and protect us, providing us with the means to survive and develop.

Social fuel needs to contain a number of attributes. By definition, it needs to be based on content that is both understood and of real interest to all participants in the conversation. It needs to provide clear frames of reference, so that any communication on the subject can work within a set and mutually agreed structure. It also needs to have a social element. Earls points out that it is social interactions around people and their relationships that have most impact and tend to be remembered best and for longer. We love talking about family and friends. Gossip is more than just idle speculation or trivial debate - it is shown to be one of the most fundamental conversational needs we have. It needn't be about the people we know, either. Earls suggests our conversations about the lives and relationships of celebrities and people in the public eye have very much the same importance in how we maintain and develop our own social relationships.

Television has always played a critical part in providing the social fuel that powers many of these conversations, and its role is unlikely to diminish. There are a number of reasons for this;

1. Television is common currency; well over 90% daily reach, and millions watching the main peak-time programmes, guarantees that TV is uniquely placed to provide topics of interest that most parties to a conversation can relate to and understand. Even if one of the parties has not seen the programme or celebrity being discussed, the coverage across other media will ensure they can at least hold an opinion based on personal experience or knowledge.

2. Television provides a window on people we can talk about (a celebrity divorce, a soap character's latest tragedy), helping to contextualise our own social issues and relationships. It is no coincidence that the vast majority of most talked-about or followed celebrities are best known for their appearances on TV. A large proportion of them will have been <u>made</u> by television.

3. Television is our most shared medium, by quite some distance, and so this social fuel is likely to spark conversations among people when they are physically together, relaxed and receptive to conversation. This is the oldest and, many social experts state, the most effective form of social interaction. However, we can also see the impact of this desire for social fuel in the significant numbers of Facebook fans, Twitter trends and online conversation about TV content, both programming and advertising. Technology is enabling more and more of these conversations to occur before, during and after the content appears on screen.

4. Television is topical, either because it reflects the news and issues of the day very quickly (and will be seen by a critical mass of our social contacts at or soon after broadcast) or because it makes the news itself through event television. Topicality increases relevance, and social fuel needs to be relevant in order to spark.

5. Television can provide 'badging' opportunities, where what people talk about or display helps to define them. It provides a form of identity. TV offers shorthand badging through cult shows, quirky characters and topics that cover our interest and passions.

In the UK, the tabloid press is more keenly aware of the importance of social fuel than anybody. Even though they have limited space compared to the more serious newspapers, their pages are regularly stuffed with content about TV; interviews with soap stars, stories of how British TV fares abroad, the latest celebrity gossip. On the day I am writing this, Britain's best-selling tabloid, The Sun, managed to fit in all of these stories into just one 48 page edition;

The X Factor latest. P1
Sun reveals winner of its Column Idol competition. P2,
Deal or No Deal's £250,000 winner. P3
Ratings blow for US X Factor. P5
Brit singer becomes YouTube sensation. P15
Downton Abbey is real-life wedding venue. P30
X Factor wannabe already a star in America. Supplement, P1
X Factor's One Direction to star in movie. Supplement, P1
Interview: Downton Abbey's Maria Doyle Kennedy. Supplement, P2
Ricky Gervais to film two sitcom pilots. Supplement, P2
ITV commissions drama Love Live from man behind The Prisoner and Lark Rise to Candleford. Supplement, P3
The Only Way Is Essex - only two days to go until new series! Supplement, P4
Channel 4's Fresh Meat beats Big Brother. Supplement, P4
New Channel 4 reality show set in gym. Supplement, P4

Even serious newspapers give significant coverage to television. Take The Guardian, for example. For years, the Media Guardian waged a relentless 'TV is dying' campaign, even as the main title featured more and more TV-based editorial through its listings, critics, columnists and news coverage. Why would they do that, unless their own readership research had shown a demand for it? The Guardian regularly devotes 4-5 pages in its G2 supplement to television shows and personalities and gives prominence to breaking TV stories in its main sections every day

Newspapers cover TV because social fuel is important; it gives our conversations meaning and enhances our ability to develop our social relationships. For many sociologists and social anthropologists, it provides the fundamental building blocks to understand ourselves, others and the world around us. Television's power as a 'we' medium is a significant contributor to the social fuel we deploy on a daily basis.

The virtual sofa – keeping us connected

Until the emergence of the social web, there had been little interest shown in these conversations; in fact, many of the theories regarding the impact of new technology tended to focus on the fragmented and personalised nature of our future media experiences.

The emergence of Web 2.0 changed the focus entirely. Suddenly, it was all about social interaction and relationships, the power of recommendation and the ubiquity and impact of word of mouth. As we could see these conversations forming via the analytics, we could better understand their importance, especially as they often correlated with our subsequent behaviour. Little wonder that, in the world of media and marketing, those trivial conversations suddenly took on a much greater meaning.

The work of Keller Fay, and others in this field, demonstrates that online is just the tip of a very large iceberg, and we have all seen the damage that icebergs can do! However, if we also add in telephone conversations and texting (far bigger than online in terms of sheer numbers of conversations and amount of time spent on them), then we can definitely see the increasing power of the virtual sofa, the ability to share our television experiences with people who are not in the same room.

The virtual sofa is becoming a very attractive place to be, and more people are hanging out there. It may be due to cultural reasons (e.g. the rapid increase in single-person households), technological ones (the availability of wi-fi connections and online devices, especially in the living room), availability issues (most of the social networks are incorporating TV successfully into their products, and vice-versa) or content-related (many broadcasters and advertisers are now making 'talkability' a key attribute of their content).

There are many examples from the TV schedules to demonstrate this point. Most broadcasters now employ social media experts, to support their overall strategy as well as to maximise the social buzz around individual programmes. At a basic level, we can see this every day in the constant in-programme directions to the show's Facebook or Twitter sites, the communities that are created to provide oxygen to the social fuel and the regular contributions from the social space into the programmes themselves.

Big events like The Olympics, The Oscars or even '*The Eurovision Song Contest*' have their audiences boosted by millions of fans participating via social networks. Special apps have been created (e.g. zeebox), many of them topping the download charts, providing ways for fans to speak to each other or share the live viewing experience via their second screens. Recommendation engines, integrated into the electronic programme guides, provide viewing options that we might otherwise have missed. The beauty of the networks using We-TV processes is that not only does it increase the aggregate audiences to the shows; it also persuades more of them to view live, to enjoy that shared experience.

The kinds of programmes that the virtual sofa seems made for include those with ongoing character development (e.g. soaps); those that appeal particularly to men or women (sport and/ or reality programmes generate a great deal of chat as partners in the household may be less inclined to share such experiences);

youth-oriented programmes and event television.

The BBC was particularly quick to pick up on this opportunity and are now benefitting hugely. As Ray Snoddy recently pointed out;

> *"While not claiming to know all the answers about the future, John Smith, chief executive of BBC Worldwide, the commercial arm of the corporation, is a great believer in the new opportunities the internet and social networks make available for distributing and expanding the impact of quality content. Smith is a great fan of YouTube - 800 million views of BBC clips. Apps are equally interesting with more than nine million apps for everything from BBC News and Lonely Planet and Good Food. And as for Facebook, Smith is salivating.*
>
> *Top Gear pages on Facebook were attracting around 500,000 followers. The BBC took over the curation of the pages, added more content and now has 14 million Top Gear followers with another 13,000 joining every day. Many then link into the BBC's topgear.com, from which money is made.*
>
> *"If we are smart, if we are willing to experiment, we can find astonishing new ways to get our material to fans - and still make money out of it," Smith told IBC.*
>
> *The relationship between spiders and flies will always be problematic but in the case of the television industry and the social media giants there really may be a way to co-exist, co-operate and prosper".[39]*

The power of social media to boost viewing and engagement with TV content is also a great opportunity for advertisers, and there have been a number of campaigns recently that have embraced the potential for creating dialogue among their target audiences around their advertising.

Take comparethemarket.com. The challenge for a relatively unknown brand in a new, low-interest category (i.e. insurance price comparison sites) was immense, with bigger brands such as confused.com and moneysupermarket.com generating much higher shares of market. Most advertising in this sector had been simple, repetitive (mainly focussed on hammering the website address into people's long term memory) and rational - after all, what could be more rational than deciding where to compare insurance deals?

Comparethemarket.com, with their advertising agency VCCP, took a chance – as challenger brands often do – by going completely against the market norms, when they introduced their new 'spokesperson' to head up their TV campaign in January 2009.

Aleksandr Orlov, a CGI-generated Russian meerkat, was first seen imploring viewers to stop visiting his website – comparethemeerkat.com – by accident, when they had meant to go to comparethemarket.com. Over the course of several campaigns, we were introduced to his friends, his family history and his country of birth (Meerkovo). The ongoing 'story' of this cute and intriguing character has resulted in almost a million Facebook fans (a fantastic opportunity to create an ongoing dialogue with existing and potential customers), many millions of video views, huge levels of internet traffic (the 'spoof' site comparethemeerkat gets 2 million hits per month alone!), and even the best-selling book for Christmas 2010, soundly beating Tony Blair's memoirs and autobiographies of Russell Brand, Cheryl Cole and Danii Minogue. The TV campaign resulted in comparethemarket's site becoming the 4th most visited insurance website in the UK (up from 16th before the campaign launched) and a doubling of overall sales[40].

Aleksandr the Meerkat's success has demonstrated just how far the virtual sofa can enhance a highly creative and engaging campaign. More recently, campaigns have been developed that have been specifically designed to get people talking and increase the perceived fame and engagement with the brand. One example is dairy producer Yeo Valley.

As with comparethemarket.com, Yeo Valley was a small, challenger brand in yet another competitive but low interest market with relatively low brand awareness. Traditionally, brands would often address these challenges with a heavyweight advertising campaign. Yeo Valley, with its agency (BBH), tried a different approach.

The campaign was highly creative and shareable. It featured a quartet of rapping farmers eulogising rural life in the land of Yeo Valley. It wanted to create impact, and decided the best way to achieve that was to offer an engaging slice of long-form entertainment at a time when the audience would be primed to engage and share. As such, the ad only ever appeared in ITV breaks around the new series of '*The X Factor*' in 2010, taking over the whole of the first ad break. The bond was strengthened by on-air and online competitions in conjunction with the show...but the conversation it generated went much wider.

I experienced for myself just how viral this approach could be. At the birthday party of a friend's five year old, adults and children were winding down towards home time when my wife insisted we put the latest episode of 'The *X Factor*' on in the background. Of course, among the two dozen or so people present, the majority started watching and, soon after, the Yeo Valley ad appeared. The room was completely split between those who had seen the ad the previous week and those who were seeing it for the first time, and the cacophony as the former tried to explain the ad to the latter outstripped anything the assembled kids could summon up during their earlier game of 'musical chairs'.

The ad became a top trending Twitter topic, featured in millions of Facebook conversations and status updates and helped Yeo Valley achieve an immediate 15% uplift in sales, a half million more UK households buying the product, doubled sales of the featured four-pack yoghurts, well over 2 million YouTube views and a 400% increase in site traffic. Not bad for an obscure brand in a featureless market limiting its campaign airtime around one show.

Meanwhile, the virtual sofa itself shows no signs of being abandoned or wearing out any time soon. As the social networks get even more TV-friendly (and they know more than anybody how much TV drives our online conversations) and accessibility via wi-fi and connected screens grows, we can expect more conversation centred around the TV. For some, it will replace actual sharing of TV content in the same space, but for others it will simply provide an additional way of keeping TV social and maximising the shared experience. It's all about connecting with people, and TV is one of the most effective channels to maintain those connections.

Our conversations have come under the marketing microscope since the emergence of Web 2.0, and for good reason - they drive audiences to content and consumers to brands. But, talkability is more than a way of steering the herd; it can also change the nature of the beast. What we talk about creates fame - but if fame is plentiful in these media saturated times, what fame brings is in increasingly short supply. I refer, of course, to trust.

TRUST

Who or what we talk about provides a sense of what is important in our personal lives and in the world around us. The sheer amount of conversation that now takes place as one-to-one conversation, augmented with the networking power of social media, is what is really behind the emergence of the 'Age of Dialogue'.

No brand, however, can survive just on dialogue and talkability; TalkTrack estimates that we talk about 9 brands every day, and much of that is inspired by marketing; brands that simply drop their most effective marketing channels in order to fund 'dialogue' will find dialogue much harder to maintain. Even those brands that continue their above the line marketing efforts in parallel with social media will need to work hard to generate enough dialogue to create word-of-mouth reach on any meaningful scale.

Nine brand mentions a day across a population of 60 million people sounds like a lot of dialogue, but if we break it down we can see a fundamental flaw in the logic. It amounts to just under 2 billion brand mentions a year in the UK. That sounds like a lot of talk except television alone produces well over 2 billion commercial impacts every single day of the year.

Let's put it another way. More than 3,000 brands advertising on British television in 2010 alone. With 2 billion brand conversations (a significant number of which might be negative and many of which may be no more than a flippant comment) to share out between them, that averages less than 700 mentions per brand per year. Even the top mentioned brands are only likely to achieve, at best, a few hundred thousand 'impacts'; hardly enough to be considered mass reach. The majority of brands would have to survive on a diet of less than a thousand conversations about them every year. Where is the scale in that?

In fact, despite the massive investment in social media which has enabled more and more of these conversations to take place (and their impact to be measured), TalkTrack data suggests we are not talking about brands significantly more than we were in previous years. Social media have provided a slight boost, but our daily face-to-face conversations have always been the bedrock of our brand-led conversations.

Talkability, though, is not an end in itself; in fact, it appears to be best used as an indicator of the likely success of a brand's (or programme's) initial appeal, rather than its longer-term prospects, when it comes to driving consumers or viewers to a destination. However, where talkability really becomes important is in a much more implicit way - it tells us who (or what) is famous - and, as we shall see in the next chapter, fame is vital to advertisers.

If we were to ask consumers how they would define fame, it would not be based on how well-known somebody or something is to them personally, it would be based on how much they think other people are talking about them. Fame is not passive - it is active; top-of-mind rather than back-of-mind. If people aren't talking about you, you may be well-known...but you are not famous!

Fame is also something people like to cluster around. As the French philosopher Pascal wrote in the early seventeenth century;

"The charm of fame is so great that we like every object to which it is attached, even death". Fame has been seen for centuries as a guarantee of popularity. This may help explain why so many fading celebrities are willing to risk life, limb, career and dignity on TV programmes such as *'I'm A Celebrity...Get Me Out Of Here'*, *'Celebrity Big Brother'* or the dozens of more obscure programmes that are prepared to feature and humiliate them.

The same is true of brands. Whether or not people talk about the advertising they are exposed to, it can have a massive influence on whether or not they think other people are talking about the brand, which in turn will impact on the brand's appeal to them.

Brand Index, part of the YouGov group, measures attitudes to more than 2,500 brands worldwide, going back to 2007. They continuously measure brand performance against a range of metrics, including quality, satisfaction, value and recommendation. Perhaps the most important measure, and certainly the most sensitive to marketing activity, is the measure of 'buzz'; whether people have heard anything positive or negative about the brand through media or word of mouth. When Thinkbox analysed the relationship between Brand Index scores and advertising activity across three years, TV had the strongest overall relationship of any other media channel, but it was in driving the buzz metrics where TV really worked effectively. This translates directly to 'brand fame'.

This notion of fame came up spontaneously in Thinkbox's award-winning sponsorship research from 2008[41]. When respondents were asked what types of brands sponsored television programmes, comments from the focus groups included;

"Only big brands can afford to sponsor programmes"
"They must have a lot of money to be on a show like that"
"You think to yourself, 'they've done well to get on that programme", they must be a big brand"

Of all the measures of sponsorship's value that we placed into the research, it was "*this is a well-known brand*" that showed the greatest difference between viewers and non-viewers of the sponsored programmes. It makes a statement about the size, confidence and stature of the brand in question. Now, we may think that other measures are more important; awareness, recall, purchase intent or brand preference perhaps. The truth is, advertising payback is far more influenced by how much fame the brand possesses.

Within marketing, there has always been that brand leader effect; the brand that consumers believe to be the biggest and most 'famous' has a ready-made advantage over other brands in the sector, in terms of how effective and efficient their advertising is likely to be. Brand leaders' advertising tends to work harder and more efficiently because generally it is reinforcing established views about size and therefore reputation.

A similar effect is behind the popularity of celebrity endorsement of brands, another growing phenomenon. Celebrity endorsement is centuries old; in the mid nineteenth century, the originator of *haute couture,* Charles Worth, sought a high society lady and an influence on the court fashions, Princess Von Metternich (wife of the then Austria's ambassador to France and close friend of Napoleon's wife Empress Eugenie) to act as celebrity patron and supporter of *La Maison Worth*. This connection contributed immensely to the success and status of this couture house as the most influential in the world in its time.

Nowadays, all but the most gullible of people knows that these are not true endorsements. We don't expect Stacey Solomon to be serving up Iceland party snacks at her Christmas party, nor do we expect Jamie Oliver to shop for his ingredients at Sainsbury's. In many ways, it is not the endorsement that is important as much as the association. That association appears to work on two levels.

There is the relevance of the association (and part of the fun for consumers is working out the rationale behind the association between the celebrity and the brand) and there is the fame element – a sure, confident sign that the brand is big enough to be able to fritter money away on big-name celebrities (or even, in the case of brands like Iceland, a group of D-listers!)

These associations are powerful drivers of heuristics - the mental shortcuts we take to make our decision-making easier. In today's fame economy, celebrity provides mental shortcuts in more ways than one.

Fame and the peacock

Peacocks are swathed in ornamental feathers because the best display signals mating potential (most animal species have similar mating comparisons, but few are as dazzling as the peacock's). This idea that the mere act of display can determine mating success has been applied to human relationships (Neil Strauss' *'The Game: Penetrating the Secret Society of Pick-up Artists'*[42] explores this to amusing yet guilt-inducing effect) and more recently to marketing. Within branding there is a dawning realisation that fame and trust are closely linked; it is the big, famous brands we trust the most, sense have most to lose if they let us down and are perceived to be quality because...well, otherwise, how would they have become so big? It is no coincidence that display advertising has been called thus; the bigger the display, the greater the chances of 'mating' with the consumer. It is why brand leaders are so hard to shake from that position, once entrenched.

The display of the peacock's dazzling tail feathers to attract mates is one of the most visual and lucid examples of evolutionary biology. As many people in advertising[43] have recognised, the Peacock Theory can also relate to the love affair brands wish to spark with consumers. The argument runs that the more a brand advertises, or gets itself noticed (the display itself), the bigger and more famous consumers will perceive it to be. If that display is considered frivolous or even wasteful (the vibrancy of the display), then such feelings of size, fame and reputation will be reinforced. This is important to consumers (Robin Wight, of the Engine Group, refers to it as the 'reputation reflex[44]') as they feel this means the brand has a reputation, which must have been earned (an indicator of past quality and/or value) and needs to be maintained (the guarantee of future quality or value).

The Peacock Theory helps to explain the brand leader phenomenon mentioned earlier. It also helps to explain why the more creative campaigns, with the least explicit messaging within the advertising, tend to sell product more effectively. Not only is the advertising creating emotional connections to the brand, but it is also creating a sense of size, scale and brand fame; and the talkability that such advertising creates is a key element. If everybody is talking about the brand, even (especially) if it is based on something as obscure and ridiculous as, say, a drumming gorilla or rapping famers, then as far as the consumer is concerned, that brand must be famous.

The perception that fame equals reputation, and a strong reputation provides a guarantee of sorts for both past and future performance introduces an emotion that is becoming harder to elicit as consumers become more cynical and sceptical during these current political and economic times. It is something that brands are, by their very nature, designed to create. That emotion is trust.

Trust me, I'm an advertiser!

' The ultimate goal of marketing is to generate an intense bond between the consumer and the brand, and the main ingredient of this bond is trust'

Marketing Magazine, October 2007

We live in cynical times and trust was already a scarce commodity even before the present political and financial turmoil. We have seen the behaviour of bankers, business leaders, politicians, religious leaders and the media open to scrutiny in the last three or four years, and what has emerged makes it difficult for the average person to know quite where to turn to any longer. That doesn't make trust any less of an issue; quite the contrary. People want to trust, possibly more than ever before.

PR giant Edelman has been producing its Trust Barometer for eleven years now, canvassing the opinions of the 'informed' working age adults in a number of countries. The 2011 Trust Barometer provides some thought-provoking headlines.

It would be crass to say that trust is collapsing like a house of cards; after all, in many of the developing countries, trust in the main institutions of government and business is on an upward trend. In the west, though, trust is approaching record lows. In the USA, trust in all four main institutions – government, NGOs, business and media – plummeted in the last year, while in the UK trust in either business or the media to do what is right is below even the rock bottom levels which accompanied the banking crisis of 2008.

(An interesting statistic coming from the same research was that trust in technology companies was significantly higher than for the established institutions. For many people, companies like Apple, Google, Facebook, Sky, Virgin and Sony look after them far more competently and thoughtfully than the traditional institutions!)

Trust is also a cornerstone of branding; in fact, it is the very basis of branding, which was initially established to indicate a guarantee of provenance. As consumer choice increased, it has become more important in helping us through the millions of decisions we have to take every day. Brands are one of the most salient heuristics, or shortcuts, we use every day to make those choices more manageable. The thing is, as anybody who has taken a journey with a navigationally-challenged companion will know, if we take a shortcut, we have to have total trust in the source. The more shortcuts we make, the greater the trust needs to be.

Sometimes, trust has to be enforced. In the early days of the industrial revolution, the consumer was expected to take all the risk. There was even a legal phrase – *caveat emptor* (let the buyer beware) – to enshrine this in law. Over the years, though, consumers have been protected by more and more legislation to ensure they don't need to beware. Not only do the products and services they buy have to perform to set standards, but so do the marketing communications that support them.

Given the inherent belief in television's powers of seduction from its earliest days, it is not surprising that television content has been more heavily regulated than any other media channel before or since. TV has had to conform to a stricter set of rules governing honesty, truthfulness, impartiality and undue influence. At a very deep level, consumers understand this.

The Television Opinion Monitor I've referred to several times already, asked the question *"do you know if there is an organisation that checks if the advertising is legal, honest, decent and truthful?"*. Until the last time this question was asked was in 2007, TV has always been considered significantly more likely to be regulated in this way compared to print, outdoor and radio. Internet advertising is considered the least controlled of all media channels, with trust levels well under half of those recorded for television. We need to do more work in this area because, as choice grows and purchase becomes more trust-based than ever before, we need to understand how it can be achieved and, even more crucially, how it can be retained; because trust is an emotion that can take years to build and seconds to destroy.

To sum up, television creates conversation like no other medium and those conversations we have, and the marketing that prompts them much of the time, is closely associated with the whole concept of fame. If we think everybody is talking about a brand, we think it is famous, and if we think it is famous, we think it has a reputation to protect and we are much more likely to trust it. TV helps to create and reinforce that sense of trust, through our belief in the truth of what it tells us as well as the peacock's tail of creativity it provides to give that sense of scale, confidence and reputation. Greater regulation reinforces those beliefs in the consumer's mind.

So far we have explored television's strengths in what could be perceived as quite soft terms. Emotion, engagement, implicit attitudes, low involvement processing, sharing, 'buzz', fame and trust. They all seem quite remote from the hard facts of marketing, which has quite naturally prompted demands to *"show me the money!"* Fortunately, just as all of the advances in neuroscience, cognitive psychology and behavioural economics have opened our eyes to a totally different set of motivations driving the consumer, advances in statistics (especially through the development of econometrics) and research analysis have opened our eyes to how it affects the bottom line. We now know what works in advertising on an unprecedented scale and guess what? Creative, 'soft' concepts like these translate into hard cash. TV – and especially creative use of TV to generate emotions, engagement and memory – equals financial payback.

PAYBACK

In these recessionary times, everybody is focussed on payback. What is my return on investment? How quickly can I recoup my costs? How can I maximise my response?

Public service broadcasters across the world see their income under more scrutiny than ever before. The fast-growing pay TV business, reliant on subscription revenues, is also under pressure to demonstrate payback to its customers, through the value of the content that they provide.

The major debate about payback, though, has come from the advertising community. Just how effective is TV advertising compared to other marketing activities? We have always known that advertising works and the body of evidence is getting more substantial all the time. As our understanding of how advertising works increases, so do the questions about what it means in hard financial terms.

Fortunately, the breakthroughs in our understanding of advertising effectiveness have changed the way that we view payback within the whole marketing mix. Although these studies have scale, quality and credibility on their side, the most consistent attribute they hold is consistency. Even though they have come from opposite directions, they have arrived at very similar conclusions. One of those conclusions, the one that shines through every significant study conducted so far, is that television works, more than we ever realised.

Payback through the decades

The Institute of Practitioners in Advertising (IPA), the UK body representing the advertising agencies, is situated in a rather grand, early Victorian building designed by the famous architect, Thomas Cubitt. Like many buildings of that era, its offices can be slightly cramped and suffocating. On the morning of April 17th, 2007, the main presentation room was more suffocating than usual, as it was packed to the rafters for a presentation based on three decades worth of data from the IPA's Advertising Effectiveness Awards. It was, and remains, one of the most enlightening and thought-provoking presentations it has ever been my pleasure to attend!

The IPA Effectiveness Awards are considered to be the world's gold standard of advertising effectiveness and have been running annually since 1980. Each entry must provide in-depth, audited evidence to back up their submission and entries come in from all over the world. The entries are rigorously checked and evaluated by two panels of judges; having been invited onto a judging panel a couple of years ago, I can confirm the process is thorough, time-consuming and scrupulously fair.

Over their 30 year lifespan, the awards have not just provided us with fantastic insights into how and to what extent a wide range of advertising campaigns paid back on their investment, they have also accumulated a massive databank of over 1,000 case histories. In April 2007, that number was 880 and two unassuming but intellectually commanding presenters, Les Binet (who has written more award-winning papers than anybody else, before or since) and Peter Field, stood up to present the results of a rigorous meta-analysis they had conducted on all 880 papers. The presentation was called *'Marketing in the Era of Accountability'* and their aim was to show what works in advertising, and why. To say they succeeded would be an understatement!

The room positively sizzled (and not just because it was a warm day) when they announced that *"what works in advertising is not always what we think is working, and a lot of the common wisdom out there is just plain wrong"*. They then proceeded to lay into some of the shibboleths of modern marketing theory and practice with undisguised delight and in forensic detail.

Some of their findings amused and shocked me in equal measure. For example, pre-testing of commercials was shown to be a false security - pre-tested campaigns often performed less well, possibly something to do with stifling creativity. Also, the traditional communication measures that we commonly use to evaluate ad campaigns were shown to have very little relationship with actual marketing performance. It was not the established measures of awareness or recall, claimed attention or brand attribution, or even purchase intent that correlated with advertising success. How much people <u>liked</u> the advertising had much more influence on whether they would buy the product.

(At almost exactly the same time as this presentation was being given, advertising guru Paul Feldwick, one of the founders of account planning in the UK, had taken on a similar theme in demolishing 'the message myth' in his gold award-winning MRS paper)

This led to the key finding, which is still influencing changes in advertising strategy five years later. Binet and Field analysed the creative strategy of each campaign; whether it was based on one of five categories – fame (getting people talking about the brand), emotion, persuasion, information or mixed – and compared their performance. Across all 880 campaigns, it was those campaigns based on <u>fame</u> and <u>emotion</u> that were significantly more successful in terms of leading to bottom line success.

Fame campaigns were thought to have achieved their success because, according to the authors, they get the consumers to do much of their work for them; through talkability, they punch above their weight and give the impression of being bigger than they actually are. With emotional campaigns it was felt that their success was due to emotions being more effective at influencing our behaviours (as the 'Audience' chapter demonstrated) and emotional associations lasting far longer in the memory than information or rational persuasion.

Figure 2

Influence Model

Very large business effects:	fame %	emotion %	complex %	Persuasion %
Sales	58	57	45	46
Market Share	31	35	31	27
Profit	39	28	26	13
Loyalty	11	9	9	7

Source: 'Marketing in the Era of Accountability', Binet & Field, IPA 2007. Reproduced with kind permission of the IPA

In case after case, Binet and Field could demonstrate the impact of fame and emotion on financial performance, such as sales, market share, profitability and customer loyalty, even though the rational campaigns were the more dominant model. Profit in particular was interesting; the authors argued that fame-inducing and emotionally-based campaigns were far more effective at reducing price sensitivity and persuading people to pay more for the brand, reducing the need for discounting and price promotions and feeding directly into the profit levels. The power of emotion (and fame) had never been demonstrated so clearly!

So, it was no surprise when they also announced that campaigns using TV as the lead medium were far more likely to lead to business success across these key measures, even when relative spend was taken into account. Two thirds of the TV-led campaigns (66%) achieved significant business effects compared to less than half (49%) of non-TV campaigns.

Not only that, but TV was becoming <u>more</u> effective, not less. Looking at TV's pound-for-pound ability to drive market share, they found it had improved consistently throughout the 1980s, into the 1990s and right through the 2000s (and updates across the last five years suggest it is continuing to get more effective). The authors felt compelled to make a specific point about TV against a background of criticism and pessimism about its future prospects;

> *"Don't neglect TV. Far from being dead, TV advertising remains one of the most effective and efficient media. New technology and increased competition may actually be making TV more effective, not less"*
> From '*Marketing in the Era of Accountability'* Report, IPA, 2007

This last statement, for me, marks the beginning of the long journey towards TV's rehabilitation. It was the first solid (and, in my view, indisputable) evidence that TV was moving forwards not backwards as both an advertising medium and an entertainment channel. The authors attributed TV's ascent in the 1990's to increased competition from multi-channel operators leading to better targeting and lower costs per thousand. The medium's continued ascent during the first decade of the 21st century came from an external source - one that, up until recently, had been strongly associated with its demise.

Because, if television was the single, most effective medium out there, then the most effective combination of media – by far – was television combined with online. TV-led campaigns achieved a 66% 'success rate' but the equivalent figure for campaigns based around TV plus online was 73%. Who would have thought it? The grim reaper, as far as TV was concerned, was suddenly portrayed as Florence Nightingale. And, as we shall see in the following chapters, the internet's positive influence on television's future is getting stronger and more diverse.

TV advertising – 30 seconds that can last for years

They say that you don't see a bus for ages, then three come along at once. (An old friend of mine, now sadly passed away, wrote a book on statistics explaining why that should be). The same is true for major advertising effectiveness studies.

Within a couple of weeks of the IPA Study being published, another major analysis of what works and why in advertising was announced. (I need to claim an interest here; Thinkbox were the sponsors of the study).

In May 2007, Price Waterhouse Coopers presented the results of their Payback Study[45], and the comparisons with the IPA work were inevitable. Both were huge analysis exercises; the PWC study was based on 706 brands over ten to fifteen years (depending on the market) across seven different markets – hair care, fruit juice, breakfast cereals, three different car markets and insurance. Both looked at effectiveness from opposite sides of the same coin (IPA was self-selected and successful brands from a wide range of markets; PWC included all brands within a narrower range of seven markets). Both came out with results which reinforced the other while also providing additional insights.

Across the 700+ brands studied, the PWC analysis concluded that, yet again, television advertising provided the highest (and most consistent) level of payback. On average, TV investment paid back <u>four and a half times</u> in increased sales, more than 30% better than the next best performer, press. (Unfortunately, PWC couldn't include online in the analysis because the online industry couldn't provide the relevant advertiser spend data).

A number of additional findings from the PWC study also fit very easily into the paradigm offered by Binet and Field's work. For example, one of their conclusions, on realising the importance of emotion within the mix, was that *"emotional campaigns build consumer attitudes more powerfully...and more memorably"*. They are more solid and they stay with our memories longer. This helps to explain TV's dominance in payback analysis; it also helps to explain the most significant finding.

PWC's work (and much of the work in the IPA databank) is based on econometrics; using statistical analysis to identify fluctuations in business performance that can be attributed to marketing (or other) activities. Traditionally, these tend to be conducted across weeks, maybe months at best. This is, after all, enough time to analyse most campaign periods, with a few weeks either side for good measure. What the PWC study concluded was that, in TV's case at least, we ought to be looking across years.

Deliberately keeping an open mind on timescales to analyse, Andrew Sharp – PWC's Head of Brand Economics at the time – came up with the startling conclusion that, not only does TV payback more, it also continues to pay back for much, much longer than other media. For every pound of payback TV contributed in the year the advertising took place, it continued to contribute a further ninety pence the following year (and a tiny bit of residual payback the year after that!). In other words, TV's ability to influence our emotions and long-term memory produces very long-term effects.

Now, the concept of 'adstock' – a residual 'halo effect' of advertising after the campaign finishes - has been accepted by the marketing industry for several decades, but I have not seen one mention of adstock that assumes this level of longevity. Not one! A finding like this actually changes everything, because those positive effects in Year 2, based on existing evaluation systems, would not have been attributed to the TV campaign that generated them; it would have been attributed to whichever media was being used to advertise during the year in question!

I'll return to this issue of which media get the credit for which business effects later, but there was one other aspect of the PWC study that resonated with the IPA work. It was based on the thorny issue of how advertising impacts on brand value. PWC employed a research technique called conjoint analysis (also known as 'trade-off' analysis, mimicking people's purchase decisions) across the seven markets covered, to understand how much people are willing to pay for the brand name alone, compared to other factors such as price, or product attributes.

The results were again consistent with the IPA findings. Brand value was more clearly influenced by TV advertising than any other medium by far, and the brand value leaders (not necessarily the market leaders in terms of total sales, but the brands that could persuade consumers to pay more for the brand name alone) spent double the proportion of their total advertising spend on TV compared to their competitors. Even more surprisingly, what people were willing to pay for their preferred brand names – even in highly commoditised markets like motor insurance – easily eclipsed the importance they attached to relative price. (What was even more amazing was that, when PWC repeated this exercise a year later, just as the recession began to bite, brand continued to trump price in the average purchase decision, although price had increased in importance from its pre-recession levels).

Proving the power of TV – and making sure we eat our greens

Although the statistical techniques that have been developed to allow us to isolate the returns on marketing spend are tested and proven, sometimes it all seems a little bit of a black box - numbers come out to say a medium has paid back X or Y on the original investment, but how were they arrived at? What do they mean?

The only other way of demonstrating TV's payback potential would be if one could take out all of the influences other than TV spend and then observe the effects when TV advertising is introduced. The trouble is, in the highly competitive, brand-powered consumer markets we inhabit today, how would it be possible to find a market that has no other intermediary influences to complicate matters?

Canada's Television Bureau (TVB) decided to test this approach. They found a product that had no advertising, no PR, no branding and no promotion of any kind. It didn't even have a trade body. This product, despite being a super food, had no history of investment in its marketing and little impression on the minds of Canadian consumers. It was in a moribund state as far as sales were concerned and very few associations with the product were seared into the consumers' minds. That product was broccoli.

TVB created a TV campaign to push the nutritional and taste benefits of broccoli, a product that was un-loved, un-owned and yet really good for you. They scheduled an average campaign for a launch product, aiming to turn around decades of neglect in just five weeks. There was no other advertising or marketing of broccoli anywhere else across this time period. So, did it work? The answer was an unequivocal yes!

Two sets of metrics were used to define effectiveness; product sales and advertising impact. On the latter, broccoli went from being completely unmentioned to becoming the second most top-of-mind vegetable, according to IPSOS Research, while metrics such as aided awareness (over 90%), perceived nutritional value (65% higher than for other vegetables) and purchase intent rose significantly. One in eight Canadians claimed they had bought at least one extra portion of broccoli compared to their normal shop. Social media mentions went through the roof, with specific mentions of 'broccoli' or 'miracle food' increasing more than fivefold and specific search terms doubling. Most significant of all, though, was what happened to sales. In just five weeks, TV had managed to increase sales of this commoditised, low-interest, declining product by 8% over the same period the previous year; the latest data suggests the halo effect of a creative, engaging TV campaign continues.

It's not just sales, it's profitability as well

Just as I am about to put this book to bed, another payback study arrives in my inbox, again from those TV evangelists at Thinkbox. It sniffs its nose at the 1600 campaigns covered by the combined PWC and IPA research. This meta-analysis was conducted on marketing evaluation firm Ebiquity's database of more than 3,000 advertising campaigns between 2006 and 2011[46].

Using state of the art econometric analysis, Ebiquity determined that, on average, every million pounds spent on television delivers an extra £700,000 (i.e. s70 pence per pound spent) in profit. Not increased sales, but profit, after all costs of sale are subtracted. This was well above the profit per campaign achieved by any of the competitive media channels measured, although it is worth pointing out that all of them were profitable too, overall, and their ability to achieve profit was further boosted by television exposure. However, television proved to be two and a half times more effective in driving profit than all other media channels combined, as well as having a significantly greater halo effect on other brands in the portfolio. Another interesting reinforcement of the IPA findings is Ebiquity's calculation that TV had increased its cost-effectiveness over the 5 years from 2006, through lower airtime costs combined with increased profitability. That is some achievement in these challenging economic times.

In it for the long-term

As all of these payback studies suggest, the long-term impact of TV advertising has traditionally been undervalued by econometrics and other means of evaluating return on marketing investment, because the effects in years 2 and 3 have often been attributed to whichever media were active during those years. Paradoxically, there was more likelihood that TV would not feature in campaigns conducted in years 2 and 3, because the same phenomenon had convinced many advertisers that they could take time out from TV advertising with little discernible effect on their bottom line; To an extent, they were right and during times when budgets are squeezed and longer-term investment appears to be an unaffordable luxury, it is tempting just to take a year off TV advertising.

But, what goes up must come down and if a media investment is based on long-term growth rather than the quick fix of competitive media, then presumably the longer-term impact of reducing or cutting investment altogether must also be greater. So it has proven to be.

In the west, we have a reputation for not investing in the long-term economic or business opportunities; there is often a short-term approach, probably due more than anything to the dividend-led focus of the share markets. This sometimes blinds us to the bigger picture; yes, the short-term can help us optimise dividends, revenues, cash flow and profit and loss performance but there is a bigger issue at stake; brand survival!

Because the PWC analysis was conducted over a decade or more, it enabled a longer-term look at the structure of individual product markets. The importance of brands' longer-term survival was reflected in the brands that came and went, or stayed with us to become brand leaders or cash cows.

Take the breakfast cereals market for example. It's a hugely competitive market, with lots of big brand launches and regular brand extensions taking place on a monthly basis. It is also a market with an exceptionally high attrition rate. Across the thirteen years of analysis, a remarkable number of brands were launched and died; more than half launched AND died within that relatively short period. In fact, only 11% of brands survived all the way through the thirteen years. I need hardly state that those brands were much more likely to have been the big investors in TV advertising.

This issue of brand survival becomes even more important during times of economic difficulty, as we have seen since the banking crisis of 2008. During recessions, brands go bust (and, conversely, many major brands first emerge, as they can take advantage of competitors' woes <u>and</u> relatively cheap operating costs, such as advertising rates). Again, at such times, it is tempting to cut the brand-building budgets as they sometimes appear to have a less direct relationship with short-term bottom line performance; especially more emotional, longer-term media like TV.

Data2Decisions is another firm specialising in brand valuation and marketing investment analysis. As this recession began to bite, they put a large number of their clients' long-term analytics through their statistical models. What they found[47], looking specifically at TV advertising, was that the cost to a brand of cutting advertising was greater than expected. If a brand halted their TV advertising for just a year, and then resumed as normal, it would take them on average <u>five years</u> to get back to where they were. Now, five years is a lot of sales, and unfortunately this is not being calculated in terms of the opportunity cost of 'going dark' on TV when times are tough. This is another reason why more advertisers than ever before are using TV as a key part of their communications strategy, and why so many seem to come back to television just two or three years after declaring they were dropping it as an advertising medium.

How do we attribute response?

There is another issue at play when it comes to attributing return on investment to media channels. How do we evaluate and optimise direct response campaigns? Traditionally, these have been simple to analyse. The campaign would feature different response channels (PO Boxes, telephone numbers etc.) for each media channel used and then the responses through that channel would be calculated against the money spent. This worked extremely well...until the internet came along.

There is no doubt that the internet has been a brilliant response channel, and it has transformed the way response can be activated and monetised. The problem is, it has also changed the rules of response quite significantly. Suddenly the link between the prompt to respond and the channel on which the response was 'harvested' was broken. People were seeing an ad on TV, or in their local newspaper and deciding not to use the advertised phone number but to go online and respond that way. It is estimated that online response as a percentage of the total response profile quadrupled from 15% to 60% in the previous eight years[48]. Online would then be credited with the transaction, but the real influence would be unrecorded.

Not for the first time, the consumer was ahead of the experts. The traditional response analysis models showed a markedly lower cost per acquisition from online channels compared to the established media, and the money followed the models, shifting from offline to online media. But then a strange thing happened. The customer acquisition costs of online media began to significantly rise as the prompts dried up and traffic reduced.

Mediacom, the largest media agency in the UK had recognised this phenomenon earlier than most and re-jigged their planning tools, using econometrics to understand how much the offline media were contributing to the online traffic. The results across a range of brand analyses demonstrated that they had been severely underestimating the influence of offline media on online response. According to Jeremy Griffiths, their Global Business Science Officer, this more thorough analysis reduced the cost per response of all offline media, but the biggest impact was on television, whose cost per response tumbled dramatically, often comparable to those of the online channels.

Most other agencies have changed their response planning as a result, but it took several years to get there. Those were the years when one was most likely to hear the refrain *"television isn't as effective as it once was"* and that almost always related to response analysis. It also encapsulated all that was wrong with the 'TV is dead' analysis - it was based on looking at the wrong things in the wrong way, and applying binary logic (the channel that harvests the response must have been responsible for generating it) to a much more complex and fast-changing phenomenon!

Well, now we have the data and analysis tools to prove that isn't so. The role of data in all of this is paramount. Since the emergence of the internet, data has ceased to be a scarce resource and is actually growing at a rate at which we are struggling to make sense of it. The lessons learned above provide a stark lesson in the futility of having plentiful data only to analyse it in the wrong way.

Part of the problem TV has faced over the past decade or so has been that the abundant supply of data has been used like a drunk uses a lamppost; for support rather than illumination. The data has been used to further the agenda that TV was dying, that new media channels would take its place and that human behaviour was more influenced by technology than by millennia of evolutionary development.

It has also been used in a very lop-sided way; while the data was used to triumphantly acclaim the successes of the online media, it was also used to depict television and other established media in a vacuum. It was assumed that TV, print and radio would stay still and detached from the digital revolution, that technological developments would pass them by. The reality is that all these media channels are being affected themselves by these massive technological changes, creating efficiencies and transforming their role as consumer media and advertising channels. Indeed, television, the most popular and widely adopted digital medium of them all, has particularly benefited from the digital revolution, and its place at the heart of our entertainment media landscape has been cemented as a result. Its increasing ability to generate payback for those advertisers investing in it is a testament to that fact, but it didn't achieve such impressive results on its own. TV generated payback by working in conjunction with its new best friend; online.

ONLINE

TV and online – better together

In early 2007, Thinkbox commissioned a joint study with the IAB[49], looking at how TV and online worked together. I was a slightly reluctant collaborator. Despite my love of online technology, I'd begun to tire of the ceaseless sniping at the heels of TV which often came from the IAB's members. Plus we were going to do a study using online methodologies, based on a sample representing the 25% most digitally advanced UK households, about technology and how it was changing their behaviour. In my mind, the study was not designed to reveal the most optimistic picture for TV's future. I was wrong.

The study revealed an abiding love for television, even among those with the most technology and the best connectivity. I think we had lost sight of just how much of the technology early adopters had bought was often about delivering better television rather than finding replacements for it, and how they were already using the connections to enhance that television experience.

This was really apparent when the research respondents started talking about what they did while they watched TV. Laptops in the living room, with wireless access, had only recently become a reality for many of them, but they took to the experience like ducks to water (or techies to a Star Trek convention).

Binary thinking would have analysed simultaneous consumption of TV and internet from the perspective of two screens competing for attention. From the ecosystem perspective, we saw them feed off each other. Sometimes the TV or the laptop would take over; sometimes people would switch between the two. But even when people were totally engaged with what they were doing online, TV could cut through and grab their attention.

My favourite example (we filmed everything) was two young women, totally immersed in their favourite website, which they were using to plan their respective weddings, as they explained their use of it to the interviewer. It would be impossible to be more engaged. But on the TV set in the corner of the room a favourite TV ad appeared, and their focus shifted immediately, as they salivated over the (Marks & Spencer) roast chicken dinner and related how excited they always got whenever the ad came on TV. You couldn't have scripted it!

In the integration era we can stop seeing everything as a battle for attention (it is over ten years since Kevin Kelly, co-founder of Wired magazine, stated that attention was becoming the scarcest and most valuable commodity in modern communications, as we all become cognitive misers). We can look for the complementary, rather than the competitive. If we look at the relationship between TV and online from this perspective, TV's future begins to look much brighter, because online supports it, expands its reach and complements its strengths. It does this in three ways;

- as a response channel
- as a distribution channel
- as a promotional channel.

Let's take them one by one.

The new rules of response

Two thirds of the sample in the Thinkbox/IAB study said they had been online while also watching TV, most of them claiming to do it on a regular basis. They used online to check on the latest football scores, search for music they had heard on a programme, find out more about a featured celeb or simply fill in time. What really took us aback was how many of them could relate – in fine detail – whole purchase journeys they had undertaken through these two powerful communication vehicles working together. But it wasn't just what they were telling us, it was how they were telling it that was so insightful.

The research didn't just identify the phenomenon of concurrent consumption of TV and online for the first time, although this became the headline news. It provided the first research examples I'd ever seen of consumers going through the whole purchase journey <u>during the course of a single commercial break</u>. It was the sheer sense of empowerment they displayed as they talked about how they had seen something on TV (usually) and then immediately responded via online.

In my 30 years working in media research and strategy, I had never seen such an enthusiastic embrace of the benefits of advertising. It had dawned on these viewers that if they wanted to respond, they could. One man talked of hearing his teenage daughter mention an eyeliner featured on *'Gok Wan's Fashion Tips'* and secretly visiting the BBC website to buy it for her birthday. Another talked of his sudden inspiration to buy a Nintendo DS for his mum's Christmas present because he saw a middle-aged lady playing a brain training game in their TV ad. A lady talked about how she was persuaded to shop online with Boots as a result of a TV ad and now engages more with the same ads because she is just a few clicks away from responding.

Whether they were talking about home insurance, new car purchases or army recruitment, it was the combination of two very different channels working together – usually, but not always, flowing <u>from</u> TV <u>to</u> online – that turned engagement into action. With the technology for 2-screening becoming mainstream today, this sense of empowerment with advertising content is set to grow. Some of the trends we'll talk about in the final chapters – such as growing set connectivity, second screen orchestrated content, social media, mobile technology and improved TV set 'theatricality' – may already be helping people to engage with and respond to the TV advertising they watch nowadays.

Back in the late 1990s, the smart money was on red button technology to boost TV's fortunes. The belief, shared among objective analysts such as Forrester, as well as the industry itself, was that the powerful engagement TV could elicit combined with a direct and interactive response channel could create new transactional revenues and boost advertising revenues. It didn't happen, mainly because red button technology was slow, clunky, expensive to implement and – worst of all from a response generation point of view – it took viewers away from the broadcast stream, often for minutes at a time.

There were some red button successes, mainly based around either ultra-simple response mechanisms or entertaining content (the Carling Black Label 'Old Lions' campaign, featuring veteran England footballers playing for a Sunday football team, is a case in point), and many red button campaigns achieved response levels well above what they would have received via traditional direct marketing channels. Overall, though, the cost and effort involved made it less than worthwhile and interest faded during the latter half of the last decade. It also seemed to convince many in the industry that it was the model itself – responding directly to TV advertising – rather than its technological limitations that were responsible for red button's fall from grace. It was against that background that the TV + online combination almost sneaked in via the back door.

In fact, numerous studies have demonstrated that people are more than ready to respond to anything they see on TV, on their own terms. A study[50] carried out just two years after the Thinkbox/ IAB work reported 19 out of every 20 (94%) broadband users had gone online as a direct response to something they had seen on TV and increasing numbers (around two thirds) were doing it regularly.

This is the main reason why online brands are flocking to television in the UK. In 2004 a total of 34 online-only brands spent less than £10 million a year on TV. Just five years later, online brands invested over £180 million, with a total of 239 brands accounting for 5.5% of all TV spend[51] (plus the 20+ programme sponsorships in which they also invested). Online brands' TV advertising has been growing rapidly in recent years and whole new online markets have been transformed. Other media have also benefitted, but it is TV where these online brands have invested the vast bulk of their money - TV accounted for nearly three quarters of online brands' offline media spend in 2009.

The emergence of online as a response channel has completely changed the rules of response advertising. Until recently, the silo-based planning of response campaigns was based on off peak airtime (partly because call centres could not traditionally handle the call volumes generated by peak-time response levels), inexpensive production values, dedicated phone lines and specific product lines that lent themselves to these response dynamics. Now, with online becoming the dominant response channel, TV response campaigns can (and do) roll into peak airtime, the higher quality ads are seen to generate above average response levels (branding plus response!), multiple response channels are offered and any product can benefit.

Thinkbox conducted a study with Mediacom looking at how the rules of response were changing in 2009[52]. Looking at seven response campaigns across a range of markets (including finance, travel, consumer electronics and automotive), the study demonstrated just how much of the online response was generated by television advertising, despite the call to action being via dedicated telephone lines. It demonstrated how effective TV was compared to other media; TV accounted for less than a quarter of media spend but generated almost half (43%) of the campaign-driven responses. It showed how most viewers went straight to the laptop rather than bothering to telephone and how responses can vary in scale and profile depending on the type of product being advertised. In short, it showed how the old rules of response advertising need to be completely re-written, because the consumer has taken matters into their own hands and decided to respond in their own way, rather than the route suggested by the advertiser.

The rules of response are not just changing; they are affecting the full spectrum of TV advertising. How people respond, and what they respond to, is becoming bigger, better, more widespread...and more effective. It happens out of nowhere and it happens quickly – at least half of responses generated from a TV campaign now occur within five minutes of seeing the ad. As stated above, people can now go from initial awareness, through consideration, review and all the way to purchase, all within a single commercial break. That truly makes TV a point-of-sale medium and advertising revenues are finally beginning to reflect that fact. It is estimated that more than two thirds of TV advertisements now have a response mechanism and/or call to action integrated into the creative.

Martini TV – anytime, anyplace, anywhere

On the day I joined Thinkbox, the press cuttings came round and top of the pile was a double-page article in The Independent headlined *'TV? It's So Over!'*. My first instinct was to think 'what have I done? I've joined the TV trade marketing body just as a major quality newspaper is predicting its demise. Then I read the article, which talked about *"a new competitor to the dusty old box in the corner of the room"* – that competitor was mobile TV.

In fact, nothing could be <u>less</u> competitive with the dusty old (42", HD, flat-screen) box than mobile TV. If you are sitting with a big HD screen in front of you, you are hardly likely to watch TV on a mobile device instead. What that mobile device can do is take TV to new places, competing against newspapers on the daily commute, magazines on long-haul journeys or radio at our workstations. Unfortunately, though, competitive threat was the overriding narrative at the time. Much of it was based on the premise that new forms of TV, distributed in new ways, could only be disruptive and produce a downward pressure on broadcast audiences and advertising revenues. The reality has been very different.

The internet has provided the biggest shake-up to TV distribution since the launch of pay TV. It makes TV content available to an increasing number of people in a variety of different ways. Since the launch of the BBC i-player in 2007, web-based access to on demand and streamed TV content has rocketed, to the extent that the ISPs have repeatedly expressed concerns regarding the strain it puts on their current traffic capacity. In which case, they had better invest in the broadband infrastructure, because there is still a very long way to go!

More than two thirds of broadband users - more than half the total UK population - now watch on demand programming at least once a week. All of the main sites have been experiencing double digit growth every year since the i-player launched and TV on demand (TVOD) is one of the main growth areas of internet use. All of the new platforms and devices that can distribute TV (smartphones, tablets, Wi-Fi, 3G, games consoles, connected TVs etc.) will continue this trend. But, as we have already seen, it will be some time before the impact on the broadcast model will be felt. There are three reasons for this.

1. All of these new ways of consuming television still account for less than 3% of total TV viewing. Amongst the digital natives, that figure is just over 3%. Even on the latest trends, it is unlikely that it will reach more than, say, 15% before the end of the decade, at best.

2. TVOD is purely additive. We know that viewing to the linear schedule is at an all time high and all of the indications are that much of this additional on demand viewing is happening at times and in locations when previously TV viewing would not have taken place. Even when it does compete with broadcast viewing, we know that it is consolidating loyalty to the key peak-time shows among people who would otherwise probably give up on watching some series because they simply couldn't fit it in to their already busy lifestyles.

3. This is not money that is going out of the broadcast pot towards new, online-only players. Not only are we seeing consolidation of audiences around big, catch-up shows, we are also seeing consolidation around the big broadcast channel brands. Despite the plethora of competitive on demand TV providers, who garnered so much media attention before the launch of the BBC i-player, the non-broadcast operators in this market are now scrambling around for less than 5% of the total online TV market between them (excluding YouTube, whose on demand TV offering is largely gaining because of the hosted broadcaster sites they have concluded deals with, including Channel 4 and five).

Meanwhile, rather than cannibalising audiences as many broadcasters feared, on demand TV is adding new audiences to the linear schedule <u>and</u> generating significant new revenues to bolster spot and sponsorship revenues. That puts paid to the argument, aired regularly through the late 1990s onwards that channel brands will become irrelevant in an on demand future, as viewers jump straight to content. Channel brands certainly appear to be important markers in the on demand market, as viewers use them to search for content. Hulu, the US-based on demand TV service, had to re-introduce classification of programmes by channel brand, because their users relied on them so much to find and decide on what to watch in a choice-filled world.

Whoever coined the phrase '*swapping analogue dollars for digital cents*' patently didn't base this analysis on the TV market. In fact, it is an almost like-for-like comparison. According to OFCOM, online TV revenues (based on free to view, subscriptions and transactional revenues) jumped from just £5.7 million in 2005 to over £150million in 2010; one of the main contributors to online's overall revenue increase across that time period. The broadcasters themselves have benefited from increasing demand for airtime – costs per thousand are generally at a premium compared to scheduled spots and they have been able to increase the number of ads they play out during on demand views. On demand TV now closely resembles the linear schedule in terms of ad minutage viewed and revenues per viewer.

There are good reasons for this;

- Audiences are growing so they can add reach to a TV campaign.
- Audiences skew towards lighter-viewing upmarket and younger viewers – a profile that advertisers have always been prepared to pay more to reach.
- The 'pull' nature of viewing means that audiences are

much more likely to be fully engaged with the show, and the association with on demand adds a sponsorship element; the advertisers are perceived as making helping to bring the on demand viewing opportunity to the viewer.

• Viewing is only one click away from interactivity; although, as we shall see in the next chapter, pretty soon most TV viewing will only be a click away from interactivity, if it isn't already through the addition of second screens into the viewing environment.

Social TV – TV's informal PR machine

The third benefit that online brings to television is the ability to promote programmes and any other content viewers engage with, including advertising content. It is, in effect, TV's informal PR machine.

This is not surprising. As we saw in the 'Social' chapter above, TV is a hugely social experience, and always has been. It was just well below the radar for half a century or so. Now, with the advent of social media and the speed with which those shared experiences can be spread, TV is seeing immediate and significant benefits.

In the summer of 2009, the ability of social media to create giant digital water coolers out of nowhere became global knowledge. During the auditions phase of '*Britain's Got Talent*', a contestant provided one of those OMG moments that only television can do. Susan Boyle, a shy spinster from Scotland, ambled on stage in her frumpy outfit and unstyled hair. The camera panned the row of judges, who could barely contain their sniggers at the prospect of another bedroom singer creating car crash TV. When she bashfully stated that she modelled herself on Celine Dion, the sniggers turned to looks of incredulity. Then she composed herself, waited nervously for the introductory bars to play, and began to sing *'I Have a Dream'* from *'Les Miserables'*. The voice totally belied the image, and the audience went ballistic; needless to say, she got her required votes from the judges.

It was what happened next that opened people's eyes to the massive power of TV and social media working in real-time tandem. Susan Boyle pretty much by-passed the judging process and became a global celebrity within a few days of her appearance, following over two hundred million views of her performance on YouTube and other social media sites. This highly unlikely superstar went on to become the biggest selling album artist in the world, a media sensation and, for many, an inspiration.

Meanwhile, the television programme that spawned her new-found fame saw an immediate (and sustained) increase of 2 million viewers to its audience and went on to rack up a record 18 million viewers for the final (when 'Su-bo' was beaten into second place by dance group Diversity). It was the virtual sofa that got her there; the vast majority of those who had checked out her performance were moved to do so by friends, family and colleagues asking "did you see that?" and immediate access to on demand clips providing them with the means to do so. The *'Britain's Got Talent'* franchise, like many other reality shows that have adopted similar strategies, has continued to thrive ever since that tipping point moment.

A whole host of shows, from drama (*'House'*, *'NCIS'*, *Shameless* and *'Downton Abbey'*) and comedy (*'Peep Show'*, *'Two and a Half Men'*) to sport and current affairs (witness the Twitter activity which accompanied UK television's first ever televised party election debates during the 2010 general election) have demonstrated increases in ratings, loyalty and online presence as a result of this powerful combination of platforms to ignite the social fuel and spread experiences like wildfire.

As we saw earlier, advertisers are taking advantage as well. They can promote their campaigns upfront, generate buzz from the initial airings or provide a longer tail to the brand experience once the campaign is well under way, and beyond.

Cadburys have been doing this longer than most and the arresting sight of a very lifelike gorilla rocking out on drums to Phil Collins' *'Something in the Air Tonight'* set them off like a rocket.

This century old brand was fast running out of steam by 2007, with flat-lining sales, declining profitability, an ageing consumer profile and an over-reliance on price promotions. It was also reeling from a highly publicised product recall in 2006, because bars had been found to contain the salmonella bacteria. The new marketing team needed to move fast. They also needed to make an impact. To say they achieved that would be an understatement.

When the 'Cadbury's Drumming Gorilla' ad launched on 31st August, 2007, we can probably date that as advertising's Su-bo moment. The full, 90 second version played up on the anticipation with a series of close-ups as the music slowly (very slowly) builds, only revealing the crazy premise of the ad – it's a giant, lifelike gorilla...*and it's playing the drums*! – as he (she?) launched into the crashing drum intro. The merest hint of the brand comes in at the end, under the suggestive credit to 'A Glass and a Half Full Productions' , alluding to Cadbury's campaigns from previous decades, but it comes in just at the point people would be going 'WTF!' and trying to make sense of it all. The fact that they then continued to ask and answer that question (with a great deal of speculation regarding what exactly Cadbury's were trying to tell them) kept the brand top of mind...and much better loved.

The creative team behind Gorilla went on to achieve similar results with 'Eyebrows', featuring two deadpan children doing an eyebrow dance to a 1980s electro tune (Freestyle – *'Don't Stop The Rock'* if you've suddenly got that tune playing in your head again; one of the longer-term benefits of TV advertising). But it was the jaw-dropping moment at the beginning of that drum roll that changed Cadbury's fortunes and possibly changed the future of TV advertising. A campaign that had totally resonated with the IPA's insights into what works in advertising also happened to produce instant profits; almost £5 in additional sales for every £1 spent, plus the creation of long-term value for a brand which is now forever associated with joyful moments. Moments which were shared (More than 24 million YouTube views alone), engaged with (Cadbury became the second most engaged with brand on Hall & Partners 'Tracker' research, above Amazon, the BBC and Facebook) and even inspiring social participation with the brand; just witness some of the wonderful spoofs of both the 'Gorilla' and 'Eyebrows' ads on YouTube.

It also created pride within the organisation (another often-forgotten benefit of TV advertising). As Trevor Bond, Cadbury's MD in the UK, stated;

"The Gorilla did three things for us: Put a smile on our consumers' face; Put a buzz in our organisation and made us feel proud of being part of it again; made us go back to the iconic advertising we were once famous for".

There was one other feature of the 'Gorilla' campaign's success that has always been pretty unique to television; the buzz it generated led to a great deal of editorial coverage in other media. It was featured in national newspapers (including the broadsheets, such as the FT, the Independent and the Guardian), magazines (like Zoo, Heat, the NME and Kerrang), radio stations (including GWR and Kerrang Radio) and TV programmes (such as GMTV and many news programmes). The brand was part of the national culture, almost overnight, and clever seeding (or was it part of the conversation anyway?) of water cooler topics such as *'was that really Phil Collins in the monkey suit?'* and *'whatever happened to Phil Collins anyway?')* maintained the impact well beyond the campaign's reach and well after it had finished.

This interweaving of media to tell, respond to and spread the stories that are the lifeblood of human interaction is nothing new, but it is being spurred on by what technology can do now, and it urgently requires a re-thinking of how we perceive media. As I've already mentioned, it requires a move from a silo-based to an integrationist approach to media strategy and planning. But, as people receive, process and engage with these stories – whether they are about an unassuming Scottish spinster's journey to global celebrity or the 'meaning' behind a drumming Gorilla – those stories need to be stronger, clearer and more universal than ever before. In short, they need a structure revolving around a big, engaging, shareable idea that will be accessed in a non-linear way across a variety of contexts. In order to provide transmedia storytelling, they need a narrative spine.

Integration requires a narrative spine

Modern capitalism is enduring a fresh wave of protest. The economists have predicted wrongly (but then any economist will tell you there are only three types of economist, those that can count and those that can't). The markets are failing, pensions are plummeting in value and just about every role model we may have had faith in has fallen by the wayside; politicians, bankers, business leaders and the media have all disappointed expectations during one of the most turbulent times in our recent history. So, where is the outcry?

There have been many cries in the dark, but one of things I have found most baffling in the midst of all of this uncertainty, anger and protest is the lack of a single voice providing a coherent narrative around which a set of ideas and actions can develop. A narrative spine, that provides structure and meaning to what seems like a set of disparate beliefs. After all, we know how important a sense of narrative can be in our understanding of the world...and how we behave within that world.

So it is with integrated media. I think this shift to exploring how media work together and how they either complement or compete has opened our eyes to just how powerful TV has been, not just in and of itself but in terms of its effect on other media. And I'm not just talking about online media.

TV works differently to print, for example, which itself works in a variety of ways - newspaper double page spread, a magazine advertorial, a brochure or a sixteen sheet poster. They play different parts in the marketing process. But, the more media there are and the more we integrate them in media planning, the more they need to work around a central idea, a narrative spine. Without it, they become like the lone Occupy protesters; isolated, and ultimately without a coherent, universally understood meaning.

The problem with media integration, especially as we are mixing those media experiences more and more, is that every participant will have a totally different range of interactions and experiences with a brand. Gone are the days when we all consumed the same content *en masse* and the range of channels available to display that content was limited. Nowadays, marketers can't control the way we consume their advertising. So, how do they guarantee that we 'experience' the brand in a way that is meaningful, memorable and motivating?

They do two things. They ensure the narrative spine does everything that is required of it in terms of engaging consumers with the big brand idea; and they ensure that idea is reinforced, enhanced, personalised and contextualised whenever and wherever they come across it. In essence, they engage in transmedia storytelling.

From connection to convergence

So, online technology has transformed the TV landscape, and in a dramatically more positive fashion than anybody predicted. It creates response opportunities, creates more viewing options and promotes TV content through social media. It has created the connected viewer, using other devices and platforms to synchronise with and enhance their TV viewing.

But, as we progress through television's third age, viewers are not just getting connected, they are also getting converged. Digital convergence – of content, devices and experiences – is already a fact of life in many households and will become mainstream very quickly. Even the humble television set is now part of that converged and connected future, as connected or smart TVs become the norm. What will that near future look like, when TV sets can talk to other devices and TV goes everywhere (and everything comes to the TV)?

It's coming soon...to a TV set near you!

PART THREE

SOON

CONNECTED

Iburbia is located in an unprepossessing building in Chiswick, a leafy West London suburb. Anonymous from the outside, it could house anything from a graphic design studio to a kindergarten. In fact, it showcases the latest technology – especially the technology in the living room – and hosts demonstrations, technology reviews and consumer research groups to make sense of the dazzling array of kit that is available to the average household. Like painting the Forth Bridge, it is never completed.

Nigel Walley is the energetic, almost evangelical CEO of Decipher, the strategy consultancy that owns Iburbia. He looks at all new technology from the consumer perspective, rather than simply its functionality. He has run Decipher since 1998 and has found the pace of change in the last couple of years to have been both exhilarating and exhausting in equal measure. He can plot a rapid accumulation of new launches in the areas of theatricality (e.g. 3D), user interface (e.g. movement-sensitive remote controls), on demand programming (e.g. Google TV), second screen services and – perhaps most important of all – connected TVs.

Walley and his team are relentless when testing the consumer acceptability of new products and services in this arena. They have to be. Not only are there too many launches for the consumer to take in, but most of these new products and services require far too much effort from consumers who are usually dog-tired and overdosed on decision-making by the time they hit the sofa. Walley believes that many of these new products get it so wrong because they are developed by 'techies' who can easily relate to traditional online experiences based on a lean-forward approach to interactivity, but fail to understand the lean-back mindset that dominates most TV viewing situations.

Plus ca change

Walley charted the array of new solutions (many of them still looking for a problem to solve) during 2010 alone, and even he was surprised by how many potential game-changers the industry has to get to grips with and analyse in a single year. Based solely on launches that he considered would have some immediate impact – such as the launch of Sky's Anytime Plus, Sky 3D or the broadcaster on demand services on games consoles - the industry would have had to adjust for 25 different innovations reaching the market in one year alone!

The innovation can be grouped into three key product areas;

1. **Timeshift** – enabling people to watch any piece of TV content at the time of their choosing. As we have already seen, this is almost entirely additional to the existing TV audience and is helping to increase loyalty to favourite programmes.

2. **Placeshift** – enabling people to watch any TV content at the place of their choosing. TV really is becoming mobile, through devices smartphones and tablets (in addition to the ubiquity of TV screens in public places).

3. **TV Plus** – devices or services that improve the core viewing experience, whether through improved theatricality (3D, HD) or additional content, often consumed on side screens, and increasingly synchronised with the broadcast content

This has led to innovation in four main product areas, which is leading us inexorably, and rapidly, to the truly converged connected TV experience. These are;

1. Theatricality through better screens and HD/3D broadcast standards
2. TV availability via different devices
3. Web functionality merging into TV sets
4. Integration across home networks

Let's look at these one by one.

'Pure theatre' – why experience trumps convenience

During the early part of the current recession, one of the major casualties was expected to be the home electronics sector; after all, when money gets tight, then the first thing to go is those expensive leisure items, right? Not always...

In the worst (so far!) year of the recession, 2008, Samsung increased worldwide sales of flat screen TVs by 56% and all of the major electrical retailers attributed a big part of their financial performance to TV set sales . The industry reported double digit growth for each of the previous seven years. Helped by falling prices, they were reported as one of only two bright spots in the electrical market (the other being low energy light bulbs) in a BBC report[53] on the impact of the recession in February 2009.

During tough economic times, TV equipment has been one of the few items that consumers not only spent significant money to obtain, but also happily pay a premium for features that they don't even use. Most HD-ready sets are still not receiving HD programming (although many of the owners think they do) and a significant percentage of connected TV sets are still not yet connected to the internet. Both are examples of people future-proofing the technology they buy – for many it will be a while before they can buy another new set. Both are also examples of the importance of theatricality – the quality of the viewing experience in the future TV mix.

In her excellent analysis of TV's future, *'The Television Will Be Revolutionised'[54]*, Amanda D. Lotz, focuses on the importance of theatricality within the new TV landscape. Admittedly, the impact of HD in the USA has been greater than Europe, because the previous standard definition had been noticeably inferior in terms of picture quality. That said, HDTV has already significantly changed viewing behaviour and has reportedly improved engagement levels with both advertising and programme content. To quote from the book;

> *"But to those who own HD sets in the United States, there has been a radical adjustment in the visual experience of television. High definition provides such an improvement in image quality that whether or not a program is available in this format can become a determinant in selecting what to view."*

Theatricality can determine more than viewing preference. In the UK, where the difference between standard and high definition is less pronounced, BSkyB recently reported that their Sky HD service was the fastest selling new product launch in its history, hitting 300,000 homes in its first year of launch; considering that its predecessor was Sky Plus, which now resides in more than three quarters of all Sky households, that is impressive. Since 2007, HD penetration has risen from 5% to 42% and continues to grow (up 13% in the 12 months to July 2012 alone[55]).

3D has also made an impact. Considering sets are expensive, the technology is still evolving and 3D content is very limited, it is amazing that well over a million UK homes own a 3D set[56] and at least one in five of them subscribe to Sky's 3D channel, only launched a year ago.

All of this helps to explain why television has remained so strong. We want the best viewing experience possible, we're prepared to pay good money for it, and we will make huge compromises – including which programmes we select to watch on which screen.

One final point about theatricality; it increases TV's impact. Back in 2007, when HD was just starting to take off, respected US media analyst Jack Myers produced research[57] to show viewers with HD sets were *"10 per cent more engaged"* with ads on HD channels, even though all of the ads were transmitted in standard definition at that time. More recently[58], researchers from the University of Austin, Texas, were able to conclude;

"Viewing in HD on a large screen positively impacts attitude and purchase intent".

Now, I don't agree with research that asks people how engaged they are, I believe they are too immersed in engaging content to self-report it accurately BUT better picture quality, bigger screens and more dimensions can only add to their engagement and, as we have already seen, engagement is good for both programmes and brands. It's about looking (and sounding) good.

TV availability - anywhere...and everywhere

We are becoming surrounded by screens, almost as if *'Minority Report'* has arrived several years early. In fact, many of the technologies featured in that film are already with us[59]. Audio visual is fast becoming the default communications vehicle, as 'rich' media invade our lives. Television is a central, if not the central organism in that thriving AV ecosystem.

How different to the scene described by the doom-mongers of just a couple of years ago. The excitement they felt about broadband's rich media potential shows they believe in the power of AV, but its emerging ubiquity was perceived to be a problem for the TV industry. If online could deliver rich media, it would create the disruption already experienced in print and sound based industries, they confidently predicted. Surely people would use these innovations to actively avoid TV and explore alternatives. Wouldn't they?

The reality has turned out quite different. Total consumption of video content has rocketed on the web as people hunt down viral videos, advertisers demonstrate product features using AV, user-generated content is increasingly audio-visual and news providers invest in TV-like content to bring their services to life. But none of this has been at the expense of TV viewing levels and much of it goes to support and enrich the television experience.

It is incredible to think just how much time we spend with screens these days. Not only our TV sets and tablets, laptops and smartphones, but also cinema screens, screens in bars and pubs, digital posters and even TVs in the back of taxi cabs and people carriers. Screens are everywhere, providing a constant backdrop to our lives. The world is becoming an audio visual place.

This not only reinforces TV's leading role in modern communications, including advertising, it also means TV-like content needs to be cut, expanded, mixed or mashed to accommodate all of these different contexts and mindsets. The positive side of that is once you've invested in TV content; it can go a great deal further than the TV screen.

Liverpool Street station in London's financial heart is not an especially exciting place to be on a cold January morning, but at 11am on Thursday 15th January 2010, the usual collection of commuters, shoppers and tourists were in for a shock. The distinctive opening vocal from Lulu's 1960s version of '*Shout*' started playing across the PA and a few people dotted across the concourse began to dance to it. The music segued into the Yazz classic '*The Only Way Is Up*', the Pussycat Dolls asking if we thought our girlfriends were hot like them and, bizarrely, '*The Blue Danube*'. More and more well choreographed dancers (as we began to realise this was a stunt) got up and boogied, through Kool & the Gang, '*My Boy Lollipop*' and The Beatle's '*Twist*'. Just as suddenly, the 350 participants in possibly the most filmed and shared flash mob of all time walked off in unison and challenger mobile operator T-Mobile had a great ad to reinforce their 'Life is for sharing' brand communication.

The ad took over a break on '*Big Brother*' the following evening and then seemed to take over our lives. It received more than seven million viral video views, topped the Google charts for advertiser search and was covered extensively via mainstream media – several TV programmes even attempted to recreate the dance. It featured across all T-mobile's communications channels, attracting dozens of Facebook fan pages and hundreds of thousands of Facebook fans.

Most importantly, though, the T-Mobile ad offered a new approach to television advertising. Deliberately blurring the boundaries between TV advertising, event marketing, PR and branded content, the company was able to create well over a hundred different content properties from the one piece of event filming, from direct response ads to digital outdoor, from interactive TV (it has become one of the top ten most successful iTV campaigns ever) to user-generated content. 'Dance' seemed to surround UK culture for several months and has since spawned similarly successful content launches (who could forget their dance spoof of the royal wedding in April 2011?), all emanating from one piece of brilliant audio visual content.

So, AV is everywhere – at work, rest and play, to quote a famous Mars slogan. TV content – programmes, ads, extended content and clips – plays a huge part in that. Increasingly, TV screens are everywhere and everything is on our TVs.

TV is everywhere because of the massive increase in the devices that can play it back. Smartphones are now more prevalent than traditional mobiles and tablets are becoming increasingly common, with more than 10% of the UK population owning one (up from around 3% a year ago). These provide both additional viewing opportunities, often in situations where TV viewing has traditionally not been possible, and enhanced viewing experiences when they are used as a companion screen. Advertisers are responding both creatively and strategically. The opportunities to add value to the TV advertising proposition are about to multiply - because TVs are getting connected.

Connected TV...but not 'Web-TV'!

The web TV experience envisaged by the likes of Microsoft spectacularly failed to ignite the public's imagination, because it failed to understand the immersive nature of TV viewing and attempted to tack a web experience onto a TV one. Viewers generally found it to be distracting, clunky and largely irrelevant... because they were there to watch television!

More recently, interactive television – otherwise known as the 'red button' – has attracted similar criticism. Red button interactivity works well when offering additional, immersive TV content (e.g. coverage of all Champions League games, when only one can be shown on the main channel) and for some as a separate internet-lite information service, but when it tried to gatecrash the viewing experience, by taking viewers away from what they were watching, results were generally disappointing and many advertisers decided it was more trouble (and more expensive) than it was worth.

As we have seen, web functionality is quickly merging into TV sets (as well as sitting alongside them) as more people purchase connected TVs. Many second screen activities –instant response, access to content, sharing our experiences – can now potentially be conducted directly via the connected TV screen. But will connected TV sets be used in that way? Just because they can doesn't mean they will, and the numerous players who wish to move into this space (broadcasters, platforms, console manufacturers, set manufacturers, middleware providers and advertisers) will need to think long and hard about how they are going to cut through with genuine value added services that require the minimum of effort on the audience's part, against a backdrop of competitive noise and consumer inertia.

I was asked to explore this issue when connected TVs were just a speck on the horizon and we had little direct evidence on how consumers would use them. Having analysed the failure of Web TV and red button, as well as the dramatic impact of two-screen viewing, I wrote a paper called *'One screen or two?'* and I've been delighted at how closely the current market reflects my predictions.

What we do purely via the main, connected TV screen and what we do via second or third screens will depend on a number of factors, such as context, availability and mindset. In general, though, we can expect to see the following;

- **One screen** activities – where all transactions are conducted via the connected TV set – will be primarily reserved either for access to more TV-like content, whether streamed or on demand (via broadcaster on demand services, retailers like Lovefilm or Netflix or aggregators like YouTube) or for very simple, non-intrusive 'impulse' response, which does not need to intrude too much into the viewing flow. There may be some opportunities to merge content-related chat into the viewing (e.g. Twitter feeds, closed groups of friends and fellow fans etc.) but most personal social networking still belongs on a private screen

- **Multi screen** activities will be reserved for deeper response, accessing content that would normally require most, if not all, of the screen. This includes search, research, information and product purchase. Second screens will also be the main source for accessing non-TV content that may be related to what is on screen (e.g. finding the name of the song featured in an ad or the brand of clothing a model is wearing). Finally, most social activities will take place on the second, personal screen, rather than the shared TV screen.

One intriguing way of enhancing the two screen experience could be through applications such as zeebox. Fronted by Anthony Rose, the visionary, wild-haired ex-BBC executive (he helped launch the iplayer), the zeebox app is designed as a remote control, navigation and recommendations device, social networking companion and interactive interface...all linked to the programme that is being watched on TV.

Far from being 'disruptive' (which is a favourite phrase of technology providers and consultants) such applications instead offer a consolidation of the core viewing experience, enhancing what the viewer is already there to enjoy as well as continually offering manageable choice, thereby creating more potential viewing occasions. Meanwhile, it leaves the screen to do what it does best - display immersive, rich media audio-visual content. The viewer calls it television.

Early consumer testing has been very positive and I can see zeebox, and apps like it, enjoying increasing adoption as smart portable devices become mainstream.

Either way, connected TV viewers of the future will have unparalleled power to orchestrate their own media experiences, whether through one screen or several. This is not the same as the fragmented, personalised landscape that had been predicted. We know that much of the motivation for these connected activities (and let's bear in mind that it will still be a minority activity for the foreseeable future) will come from social influences. As such, rather than look at this from a fragmentation – consolidation paradigm, we should see it as a mix of the two, consolidation of audiences but a fragmentation of the myriad of individual options with which they can enjoy that increasingly shared content in their own preferred ways.

Home networking - screen shall speak unto screen

Almost a decade ago, I presented the Living Health interactive health proposition to a team of senior public healthcare professionals. Despite its stellar performance in its pilot phase and the awards it had garnered, the IT team in the room hated the fact it had been designed to be delivered by digital TV. I was flabbergasted when, at the end of the presentation, one of them piped up that *"it doesn't matter whether it is available via TV or the web because in five years every home will have an integrated media network to make these distinctions irrelevant".* I'm still waiting.

In fact, the integrated media experience is (finally) with many of us, but not in the way that was envisaged. Rather than some central device linking all of our content together, so that we can consume anything anywhere, we have gradually accumulated separate devices that can do that job independently of a separate media centre. The final piece in that jigsaw will be the connectivity that will be standard for the majority of new TV sets and is already the case for many pay TV subscribers. Once that connectivity is in place within the home, we will begin to see the convergence that was promised more than a decade ago.

It was often assumed convergence was about getting access to any content on any screen to suit our own, personal requirements. 'Martini TV' was often cited as a major consumer benefit of convergence. But, as screens integrate and communicate with each other via the home's broadband wireless networks, it is taking on a totally different meaning. Convergence is not just any content on any screen, but screens talking to each other, making consumer experiences more seamless, convenient, richer, faster and more effective. We are already beginning to experience the first stirrings of the real convergence revolution and, yet again, TV appears to come out of it rather well. Recent campaigns for advertisers such as Honda have allowed viewers to 'capture' content from one screen onto another and retailers are piloting technology that would allow brands featured on TV ads to be automatically downloaded into their customers' shopping baskets by a simple wave of their smartphones!

In the late 1990s, Forrester Research coined the term 'lazy interactivity' to describe the way that transactional television could open up new ways for audiences to engage with television and broadcasters to unleash new revenue streams. In the final analysis the technology could never match the promise, but the concept lives on, and the ability of the different devices to do a lot of the work for us means we're more likely to engage in lazy interactivity across a range of screens. 'Just make it easy for us', pleads the audience, and now we're beginning to see <u>real</u> convergence – screen speaking to screen – make that possible. The benefits for both the viewer and the advertiser are truly exciting.

TELLYPORTING

At about the time 'lazy interactivity' was first identified, I was in charge of research and channel development at Flextech TV. At the time, it felt like we were in uncharted territory, exploring the types of content and services that would be possible with the impending arrival of iTV, 3G mobile and wireless broadband. I had to help determine the best synergies of these new distribution and interactive channels to deliver content and services.

It would have been pointless to recruit focus groups or survey respondents to answer questions based on such abstract concepts. Rather, I stood by one simple principle; "*if you want to ask people about the future, you have to take them there first*'. Once you produce an experience, and asked people to interact with it, they are far more able to explain and project their future behaviour. So, we created the experience as realistically as we could, exploring what it offered and how it could improve consumers' lives. It worked every time.

More recently, as I was grappling with the current wave of technological change, I worked with Nigel Walley and his team at Decipher on a research project that took this principle one step further and watched what happened when mainstream families are transported to the future for several weeks.

The 'Tellyporting' research was my final project at Thinkbox and in many ways the final piece of the jigsaw. We had already observed how people engage with TV (through the Engagement Study), how it paid back its investment and how the new online technologies were transforming (and enhancing) its role, as both an entertainment medium and an advertising channel. What Tellyporting allowed us to do was provide a glimpse into TV's near future - as clear a glimpse as we could get.

We placed the latest technology (including connected TVs, IPTV services, app-laden smartphones, slingboxes and – where possible – 3D television sets with subscriptions) with a range of mainstream families and compared their behaviour with existing early adopters. We also ran quantitative research to provide a benchmark of market behaviour as well as expert interviews to provide both a sense of perspective and insights for the future.

We left the equipment with the families for six weeks and conducted depth interviews before, during and after that period. What we found was sometimes predictable and often surprising.

On one level, we were disappointed. There was little contrast between the behaviour of early adopters and 'tellyported' mainstreamers. The mainstream families quickly adapted their behaviour (although, to be fair, they did get in-home demonstrations on how to use the equipment - most of us don't get that level of support) and settled into a viewing groove that was remarkably similar to the early adopters. That shouldn't have surprised us; early adopters of TV equipment are not always typical of early adopters of other technology.

Both groups tended to see these new forms of viewing as fitting around (not replacing) the TV schedules. They tended to see on demand as a 'safety net' rather than a 'destination' (which was much more likely to be the PVR); and they were more interested in theatricality (e.g. 3D television) than connectivity.

On another level, the scale and ambition of the research, together with Decipher's expert input, provided us with five key insights into how viewers are likely to adapt to TV's connected near future, all of which are interconnected. The five key insights are;

1. Consolidation of audiences
2. Hierarchy of screens
3. How TV's getting social
4. The drive to live viewing
5. The commercial contract

Here's what we found out.

Consolidation through aggregation

It seems strange, even now, to be talking about consolidation within the television business when all we have heard for the past decade is *"fragmentation, fragmentation, fragmentation"*.
Fragmentation has not been the monster that had been predicted. It has produced the efficiencies that have led to TV becoming even more targetable and cost-effective. It has attracted new viewers to television, and kept them with the schedule. It has created more viewing opportunities, more engagement and more niche targeting. In short, it's had a bad press.

There appears to be a trend back towards consolidation - audiences re-aggregating around appointment-to-view content, whether that is peak-time blockbusters or instant cult classics. Timeshift and placeshift technologies are enabling people to fit more TV viewing into their increasingly busy and exhausting lives. It's no surprise that the audience for both timeshift and placeshift is significantly younger, more upmarket and lighter viewing than the average citizen. What is a surprise is that, with ever increasing choice (perhaps too much choice), they are consolidating their decisions around fewer options, both in terms of where they go and what they choose to watch.

Firstly, they are exploring less and playing catch-up more. Catch up now accounts for almost 90% of all on demand viewing, compared to less than 80% just three years before. The message is clear; there is hardly enough time to watch all our favourite programmes to search for new things to watch.

We are also seeing consolidation in terms of the on demand destinations. Broadcaster-owned destinations were responsible for the vast bulk of on demand views, either directly from their own sites or through the branded channels on YouTube.

This has led to the decimation of the third party aggregators (with the exception of YouTube, who appear to be moving as close to the broadcasters as they can get, through hosted channels etc.). Considering the acres of newsprint that were devoted to services such as Joost, Zattoo, and Blinkbox, before the BBC launched i-player in 2007, they are now very much at the margins (or extending their business model elsewhere), with a combined reach of less than 10% of the on demand audience.

There is a lesson here about the power of channel brands. Since the launch of multi-channel, it had been predicted that channel brands would be too fragile and unformed in the consumer's mind to hold their own in this brave new world, but all of the evidence is that the on demand audience knows exactly which channels schedule their favourite shows, and are quite happy to use them as short-cuts to the content. This certainly came through in the Tellyporting results.

So, audiences are consolidating again around the big shows and peak-time schedules, even when they are catching up outside peak time. They are continuing to use the broadcasters as their prime destination, even when aggregated content is available from more sources than ever before. They are moving effortlessly from screen to screen, but the trend for more of this content to appear on the TV set (which is where viewers want to see it) does not mean we should ignore the other screens that are scattered around the living-room. As our next theme shows, they are all playing their part as well.

The hierarchy of screens

The second lesson Tellyporting taught us was how TV viewers are rapidly adapting to the plethora of screens available to them. Even mainstream families quickly got to grips with the available technology and found ways of using the different devices to consume, respond to and play with their living room entertainment as they saw fit. There is no one size fits all approach to understanding how they did it; each family had their own dynamics and preferences they brought to the experience. But, one thing they all had in common was a clear sense of the hierarchy of the screens - and the content – in terms of their functionality and importance.

The television set's role as the preferred destination for immersive entertainment was reinforced and the 'tellyported' families could only see this continuing, as bigger screens and HD/3D programming became more available.

Viewers of the future don't want anything on screen which is either intrusive or irrelevant; there are other screens to handle that. Instead, they stressed that the TV set is central to their engagement (but not exclusively of their attention), so it should be uncluttered with information or content that could be handled elsewhere - unless they choose it to be there, in which case it should be simple to access and switch in and out, as they wished.

The laptop still retains its place as the preferred response channel, especially when response needs to be open-ended and/or complex. Many viewers are already in the habit of keeping their laptop ready for action when they watch TV, flitting effortlessly between the two. There are signs, though, that the laptop is beginning to lose its predominant status as the second screen, with both smartphones and tablets making inroads. Laptops are still in more homes than either of these devices – for now – so this may be the start of a transition period.

The smartphone has quickly established its niche as the prime communications vehicle, for quick sharing and social media chat. Its highly personal and portable nature means it is the place for most social activities. When offered the opportunity to migrate some of that social media content to the big screen, the general reaction was one of horror. The big screen is far too public and immersive for most personal social networking activities.

But just as this begins to settle into a nice, neat threesome, along comes the tablet.

Although tablet penetration is still only just over 10% at the time of writing, even those who have never experienced using one appear to grasp really quickly and easily how they could be used. The emerging relationship between the tablet and the TV set is a really interesting one. The tablet could have been designed for lazy interactivity and the web-lite access to content via apps seems perfect for that relationship.

Consequently, the tablet appears poised to play an important role in the emerging screen hierarchy. Sometimes it will be used to view AV content as a second TV set (it was interesting that the rationale some consumers gave for acquiring one was as a replacement TV set). Other times it acts as a social networking device, whether via Facebook, Twitter, email or Skype. Other times still, it is a response channel, although open response, other than via an app, is still likely to be left to the laptop or desktop. However, the effectiveness of apps for a touch-screen environment means that the tablet is moving more towards the space previously dominated by laptops.

There is also a hierarchy in place regarding where and how

content is accessed. The live transmission is still considered the preferable way of watching, especially for favourite programmes, but if they have been missed, then there is a clear preference for storing them on the PVR compared to watching on demand. For many, on demand feels like a safety net – a much valued one – but those who also have a PVR tended to feel that access that way was easier, the programme wouldn't disappear after a given number of days <u>and</u> it felt like the next best thing to physical ownership, for as long as they chose to store it on their planner. This indicates a human need which cloud computing may need to address at some future date.

The preference for live viewing may confuse some readers; after all, we regularly hear about people who deliberately start viewing favourite content on commercial channels ten minutes late so that they can miss the ads. As we have already seen, the desire to avoid advertising is consistently overstated, but more often than not the desire to be 'in the moment', preferably at the same time as millions of others are sharing it, is a powerful motivator. For some, it is often a necessity, driven in large part by the next theme we uncovered - the increasingly close relationship between TV and social media.

TV – a very social medium

We have already explored the shared and social nature of television. It has been a major provider of social currency and shared experience for half a century and Web 2.0 seems to have turbo-charged this aspect of TV viewing as it expands from the living room into the virtual space. The Tellyporting research demonstrated clearly that the majority of TV viewing in the near future will have a social element attached - and that this will drive viewers to television. After all, as the social media theorists have argued repeatedly, what other people are talking about influences us deeply.

We see examples every day in Twitter trends, buzz metrics, viral views, Facebook pages and fans' forums. We know it's nothing new. We've always loved talking about TV – we discuss it more than any topic apart from family and friends, it is our most shared medium and it generates acres of coverage in other media. Of course it provides the conversational glue of our social networks.

But the relationship between TV and social media is getting stronger. More screens equals more chat, especially when that chat is free, immediate and synched in with the TV viewing experience.

I won't labour this subject, I've discussed it enough already, but social TV is becoming mainstream; every month around 1 in 5 people claim to share TV content on their social networks; a similar number will chat online about the TV they have enjoyed and 1 in 10 will sign up to become a fan of at least one programme or advertiser.

The growth of smartphones and tablets is cementing this relationship and allowing for more casual and instinctive conversations. Tellyporters talked about using Twitter feeds or Facebook status updates as a source of 'company' when watching alone or with a family of less-interested viewers (e.g. watching *'Next Top Model'* in an otherwise all-male household)

And one other point about this burgeoning relationship – it is driving the audience back to live TV.

The drive to live television viewing

As with consolidation, I never expected to see the day when I would be talking about a drive back to live television viewing. So many experts argued that we were moving from live to timeshift, from push to pull, from the restrictions of the broadcast schedules to the unfettered freedom of 'me-TV'. That should teach me not to listen to experts.

We have already seen many examples of new technology running counter to our expectations and supporting the status quo rather than becoming the predicted agents of disruption. We have seen how PVRs, the ultimate ad avoiding technology, increase the overall audience for TV ads and on demand provides the benefits of catch-up rather than lots of new programme content to discover and explore. In each case, greater access to programme content has increased viewing levels overall (which none of the experts predicted) and has been well and truly absorbed into the existing ecosystem rather than showing any signs of replacing it.

The impact of social media has been different. It is not about providing access to content in new ways, although social networks like Facebook and Google + are aiming to influence what we watch directly via our friends' preferences and recommendations. Instead, it is about enabling and extending something that has always been part of TV's fundamental attraction, allowing us to communicate and share the experience however we wish. Some enterprising individuals synchronise their timeshift viewing of favourite shows so that they can experience them together, but for most this shared experience will be via the live broadcast schedules. We are seeing the impact in the record audiences watching live TV across the developed world. People want to share, they want to be 'in the moment' and they want to watch without knowing the outcome. This last point is becoming a key part of the 'drive to live' - more programmes have a clear outcome (e.g. results) and part of the fun is watching it unfold in real time.

The broadcasters are responding, not least through the development of the genres such as reality, sport, event TV and live performance that requires the audience to consume the content 'fresh'. These are the programmes that we know generate the most chat. The Tellyporting research found that viewers are finding it difficult to enjoy these programmes in timeshift mode any longer because their enjoyment is marred if the know the results and they are unable to join in with the conversations when they are most active, during or just following the live transmission.

Live TV is going to play a major role in this mixed, mashed and meshed world for some time to come. It's not just the pulling power of social media, considerable though it is, but also the increasing theatricality of things like HD and 3D, as well as the increase of synchronous content – such as the '*X Factor*' iphone app – which makes being present at the time of viewing ever more important. That one simple fact should be enough to persuade us that we are not seeing the death of television, the 30 second spot, schedulers, channel brands or appointments to view any time soon. As long as people can gather together to watch, engage, chat, share and play together, they will. The key word in that sentence is <u>together</u>.

So far, so good - audiences stay loyal to the schedules and the broadcasters. They use the plethora of screens available to enhance the experience they already know and love, and they use them in increasingly sophisticated ways. The perfect match of TV and social media – online OR offline – is driving traffic to both in unprecedented numbers and bringing the audience back to the main event, the live broadcast. But how does that affect the advertising market? Does it change the very nature of the commercial contract between television and its audience? The answer to that is yes, and in a good way.

The commercial contract

The final theme of the Tellyporting research is the way in which these technologies are changing the nature of the 'commercial contract' between viewers, advertisers and broadcasters. This opens up so many issues around the very future of advertising itself, that I will restrict myself in this section to offering a background - what I mean by the phrase 'commercial contract' and how it is perceived in general terms by the consumer. The issues it raises are so profound that I've devoted the whole of the 'Advertising' chapter to them.

The phrase 'commercial contract', like many of the more intelligent phrases I regularly use, initially came out of the mouth of a research respondent. It was during a depth interview with a Sky Plus convert, who was explaining how his new-found power to control the advertising he viewed made him more aware of that commercial contract; how the ads paid for the programmes and were appreciated as such, but also what advertisers still had to do in order for him to "let them in". What gave the phrase such resonance was its reflection of a more general view, a sense that the technologies that gave consumers such control might also result in more relevant and entertaining advertising. It was only later that we thought to ask the question 'how could that work?' Tellyporting allowed us, and the tellyported families, to not only ask the question, but to see for ourselves.

Part of the commercial contract is that viewers expect to see less advertising in general, but more which they value. They expect to be able to respond to anything that engages them and they expect the response mechanics to become simpler and more intuitive over time. They expect their responses to yield rewarding outcomes and not to hold any unforeseen consequences (privacy and junk communications were their major concerns). In short, they have a more positive and tolerant attitude to advertising, just so long as advertising performs its side of the contract.

The Tellyporting research suggested that the '30 second spot' is going to be with us for some time to come, but that it will be less of an end in itself and more of a key element in an integrated campaign. It will be primarily the kick-starter, communicating ideas and associations about the brand, telling the story, setting the tone and landscape around which all other elements of the campaign will coalesce. Viewers will become increasingly able to take the next steps on that 'brand journey' as they begin to understand the different interactions and responses that will be open to them. Based on this research, it won't be a difficult process to get them there.

ENTERTAINMENT

People have always looked to television to entertain first and foremost, more than for any other medium. (Incidentally, I think TV also does a great job of informing and educating – for example, TV news is the most popular source of news information and documentaries have been known to inspire people to make life-changing decisions).

In survey after survey, entertainment comes out top of the reasons why people say they watch TV. It is the driving force in TV consumption. In recent years, especially since it became possible to distribute audio visual content via broadband, the web has also seen its role as an entertainment platform increase, but it is still perceived as primarily a response/information/communications tool, and much of its emergence as an entertainment destination is driven by TV-related content.

Television, because of its immediacy, lack of cost and easy access is also the medium we are most likely to turn to when we are bored, want to relax or just need to fill some time. For many, that is a sign of TV's weakness - it only fills a gap until something better comes along, that it is a 'soft' destination. There may be some truth to that, although the counter argument would be that we seem to have quite a lot of time to fill and there are always other options around. So, maybe television is just uniquely equipped to satisfy our constant need for entertainment, freshness and emotionally engaging storytelling. TV viewing may not be an active behaviour, but it is usually an active choice.

The best way to approach TV's changing role is to ask ourselves the question; *"what do I want it to do for me when I switch it on?"* The answer will vary because of the context; you might want a five minute news update or a ten minute giggle on your way to work or a bit of background chatter while you're doing the housework. Or you might switch on at a certain time to specifically to catch a favourite programme that is an established part of your daily routine. Finally, you may just gaze in the direction of the screen, point the remote and silently plead *"entertain me!"*

This innate passivity perplexes many people, especially those industry commentators who see themselves as 'switched on' most of the time, but it has been a massive influence on TV's evolution and has helped to preserve and sustain such written-off concepts as channel brands, the linear schedule and the advertising break. They have survived in part because people want choices made easy for them, and the greater choice that is available, the more they need a trusted editor. We tend to have very clear ideas about what an ITV period drama, a Channel 4 comedy or a BBC2 documentary will be like and it provides us with the heuristics we need to navigate an otherwise unmanageable choice of content.

TV schedules also provide something else the audience values - the sense of serendipity, of stumbling across a programme (or advertisement) almost by chance, of consuming content that would normally never reach the top page of one's own personal EPG. In research groups, viewers regularly discuss content they just stumbled upon which then had a profound effect on them. Although social networks and recommendation could go some way towards providing that element of surprise, they are unlikely to ever replace the schedules as the source of occasional wonder and unexpected delight.

There are other clear roles that television plays at different times, and our choice of programming will reflect this. Sometimes, TV will be a treat, whether it is renting a DVD from Blockbuster or purchasing a Sky Box Office movie direct from the screen. At other times, it will be to have a laugh, kick back, be 'in the know', to catch up with what our friends are watching or to try something new. These are all variants on the theme of *'entertain me!'*

TV's main role as an entertainment medium is unlikely to change significantly in the next few years, but how it creates that entertainment could evolve and transform quite rapidly, through delivery, distribution and integration. As we have seen, technology is already enhancing TV's role as an entertainment medium but, as TV sets get connected and other devices transform the viewing ecosystem, the roles it can fulfil will only expand.

Just as long as the money is there to fund it.

A licence to print money?

The statement that owning a television franchise was *"a licence to print money"* was made by Lord Fleet, about his successful bid for the Scottish franchise, in 1957. At the time, the new medium was already starting to make serious advertising money and there was boundless optimism over TV's future. Fast forward to 2012, and the future is not so certain. If TV is to retain its place as the entertainment medium of choice, it must continue to attract the revenues. How will it fare?

The three ages of television has produced (in the UK at least), three significant business models. Together, these accounted for over £10bn in 2010, and were shared out as follows;

Public funding, via the licence fee	23%
Advertising revenues	26%
Subscription payments (pay TV)	44%
(Other – e.g. international programme sales).	7%

Let's take them one by one.

Public funding, via the BBC licence fee, is how UK television was initially funded and still accounts for around £2.5bn every year. Although it kept pace with advertising revenues for several decades, it has been placed under increasing pressure from successive governments, themselves responding to a growing sense among some consumers (or, as governments call them, voters) that forcing people to pay over £10 a month for four BBC television channels when they can choose to pay Sky or Virgin a similar amount for hundreds of channels, can no longer be justified. There has also been speculation that people who watch TV purely via online are exempt from the licence fee, which only covers viewing via television sets.

In fact, the BBC has been preparing for the possibility that it might have to develop alternative sources of funding (e.g. encrypted subscription) since the late 1990s. The commercial expansion of BBC Worldwide, which now sells some of the most popular programming and formats in the world, is opening up new revenue streams.

This makes the UK unique; there are publicly funded TV channels elsewhere, but these are usually either relatively under-funded (e.g. PSB in the States, ABC in Australia) and/or required to also take advertising (e.g. TVNZ in New Zealand and ARD and ZDF in Germany).

This could make the UK an unrepresentative location from which to map the future of television, but I believe the opposite holds true. The existence of the BBC has not prevented the rapid introduction of new revenue models, from advertising and subscription, but has helped to create quality benchmark in programme production values and product innovation, such as the BBC i-player, to catapult the UK to the leading edge of TV's third age. In short, the BBC has kept the industry honest, innovative and focussed.

Subscription revenues were minimal until the Sky/BSB merger created a critical mass of subscribers and a virtual monopoly in pay TV, which enabled BSkyB to invest heavily in both content (especially sports rights) and marketing from the early 1990s. As the amount of money spent on pay TV in the UK is now around £5bn a year, and more than half of all UK households now subscribe to pay TV services, it is worth reflecting on how far the pay TV industry has travelled since those early days.

There had always been a hard core of households who wanted more - even the highly limited pay TV services of the 1980s managed to attract tens of thousands of subscribers. What the formation of BSkyB achieved was to provide a content offering that was more attractive to a mainstream audience of tens of millions as well as the marketing clout to convert that interest into revenues.

In the UK, pay TV revenues have grown significantly and consistently. The major threats on the horizon are cord-cutting (where people leave pay TV because they can get most of the content for free online) and premium content rights. The two are tied together; it is the premium rights, especially to sporting events, that may be the final barrier to cutting the cord, although cord-cutting is still a phenomenon that is more talked about than observed at the moment. Meanwhile, the pay TV operators are beginning to add value through convenience and packaging.

Since the late 1990's, as pay TV grew from a couple of dozen channels to several hundred, the surfeit of choice started to become a disadvantage. Often subscribers would see lots of channels they didn't want to watch on their EPG and wonder why they were paying to access them. It became evident that simply providing more choice was not enough to convince subscribers of value for money. Consequently, the focus in the last 10 years has been about providing convenience in terms of both access to programming and overall packaging.

Possibly the biggest breakthrough in this transformation was the launch of Sky Plus in September 2001. Although early forms of the PVR had been available in the UK for a couple of years beforehand, Sky Plus integrated with the EPG to make timeshift recording quick, simple and convenient. It became a massive hit, reaching almost 6 million subscribers (around two thirds of Sky's total subscriber base) just 8 years later. The real power of Sky Plus, though, lay not in attracting new subscribers to the platform but in convincing existing subscribers not to churn out of the service.

Since then, both BSkyB and Virgin Television have introduced new value added services, such as on demand programmes and high definition (BSkyB have also launched a 3D service, although lack of content and cost of hardware has limited take-up for now). Sky HD has been the fastest-growing new service in Sky's history and Virgin's on demand service is now accessed by the majority of its customers every month.

Triple and even quadruple play bundling of Pay TV with landline, mobile telephone and broadband connections has further enhanced the perceived value that pay TV can offer, which means average revenue per customer (ARPU) continues to rise. In the meantime, pay TV operators are investing more on original content, as they realise they need to invest in the schedules as well as the technology.

In the UK, pay TV revenues have seen double digit growth for many years and revenues increased by more than half in the five years to 2011 although most forecasts expect it to flatten out to single digit growth levels until 2016. Globally, pay TV audiences have increased consistently, currently adding over 40 million new subscriptions every year. This indicates the resilience of pay TV, even in the midst of a one of the worst recessions on record and against an increasing array of competitive alternatives. Pay TV penetration levels may not rise much beyond the current levels, at least in the developed markets, but an increase in platform offerings seem likely to keep total revenues solid during a time when most other items of leisure expenditure are under increasing strain.

Advertising revenues should have been hit hard by now, considering that, on top of the recession, all of the technologies that were expected to have fatally wounded TV and killed off TV advertising have become established.

In fact, as we have already seen, TV advertising revenues do not appear to be suffering a systemic decline. Although there was a decline in real terms during the middle part of the 2000s, the upswing in revenues experienced across the globe since 2009 has more than compensated for this and has left TV advertising in a relatively stronger place in 2012 than it was in 2005. TV is attracting more advertisers than ever, many from the online world, as they see instant response and longer term brand value providing sufficient returns on investment to keep them coming back. TV's role as an advertising medium is changing, and these changes will offer new revenue streams and new advertisers, helping to maintain advertising's role in television's future.

TV in the UK (and many other developed markets) is taking record shares of advertising revenues. The shift from display, or brand-led advertising towards classified (almost totally due to the success of search) appears to have peaked and display has been making a comeback as branding grows in importance and the returns on media investment are better understood. TV in particular is benefitting from this, due to its sudden emergence as a 'point-of-sale medium' and its ability to drive response, through search, social media and even direct purchase. Advertising revenues will never again be an automatic licence to print money, but they are looking a great deal more robust than they did just a couple of years ago.

That doesn't mean advertising revenues will never be under threat again; the surfeit of video advertising opportunities emerging from online, together with advertising's vulnerability to economic downturn, will undoubtedly make the next few years challenging, but the theory that TV advertising revenues were in systemic decline has been challenged itself by TV advertising's recent bounce-back as well as the new insights into how TV advertising really works.

Additional Revenue Streams. In addition to the three funding sources mentioned above, there are also many new revenue streams beginning to emerge. Examples include;

- **DVD/download sales:** Although sales of DVDs are in general decline (British Video Association figures shows spend on video content overall remaining stable, despite the recession[60]), there is one segment of the market that has been witnessing significant growth levels over the last few years. Sales and rentals of TV boxed sets now account for around 15% of the total DVD market, up from around 5% in 2005[61].

  As early as 2006, Mark Lawson was charting the rise of TV boxed sets on Amazon and elsewhere, despite the predictions that it would always be a marginal market. According to his article in The Guardian newspaper;

Connected Television

"The scale of the sales has surprised almost all the producers. The consensus was always that old TV shows would never have the home-ownership potential of movies because television was inherently disposable: a medium in which most of the product was played only once and, if it went out twice, viewers wrote to the Daily Mail complaining about too many repeats"

This is yet another example of the value people place on good TV content. Indeed, at the time of writing, UK drama series '*Downton Abbey*' has just broken all global records for DVD sales of TV series. This emerging revenue stream does not appear to be declining any time soon.

- **Transactional revenues:** In the early days of digital TV, when the red button was predicted to be a significant revenue opportunity, transactional revenues were expected to generate massive new revenue streams for the TV platforms. This never materialized, but the convergence of TV and online has created a massive response opportunity for TV advertisers, broadcasters and platforms.

  One especially lucrative area is likely to be sales of other entertainment products. For example, platforms such as BSkyB are extremely well-placed to take a significant slice of the movie DVD and rentals market via download or streaming, competing directly with the likes of Blockbuster, Netflix or Lovefilm in the process. They already have the relationships with the studios, the technology to make it possible and the customer service infrastructure to offer significant competitive advantages. We can expect to see more deals in place between premium content owners and TV businesses to promote and sell the content via their platforms.

Meanwhile, micro transactions, to enable easy payment for premium content or other services (for example, voting on reality shows or entering competitions) should also grow as companies like ITV, BSkyB and Virgin create the necessary infrastructure. The success of phone voting (where transactions are paid for via telephone billing systems) demonstrates the potential and companies such as ITV and Channel 4 are making micro-transactions a strategic priority.

Other transactional revenues, including merchandising, home shopping and ticketing are also likely to increase as the purchase journey becomes faster, more efficient and more cost-effective. Let's take home shopping on TV, for example. The UK TV shopping market is estimated to be worth £765m a year and has proven to be recession-resistant. According to the report from Verdict;

"In an increasingly multichannel retail world, TV channels are looking to boost sales by moving online, and big high street names including Argos and Debenhams are trying to tap into the growing appeal of TV selling".

• **Overseas programme & format sales:** Although the UK is fortunate to have English language content that will inevitably sell relatively well in other English-speaking territories, this is a growing market for broadcasters globally, especially as there are far more distribution opportunities, consumer demand (e.g. from expat communities), channels and platforms. Recent successes such as Danish police drama *'The Killing'* on BBC4 would never have been possible in television's previous ages.

- **The value of data:** Data will be a vital currency in this new world of dialogue and participation and television is a perfect vehicle to persuade people to provide it; after all, attractive content, trusted providers, a shared experience and a one-click response mechanism provide a pretty powerful incentive. In addition, TV will reach the light data providers; those who do not leave a joined-up digital data trail and need plenty of persuading in terms of providing details about themselves.

So, overall, the funding prospects for television look good, and are growing across multiple revenue streams. Given the resilience of the advertising market, the value people place on their television experiences and the emerging revenue streams, TV should have the revenues and the confidence to continue to invest in top-quality audio-visual entertainment. As with cinema, revenues will come from many of the sources that were meant to be providing the competitive threat - and as with cinema, this will eventually result in a very different entertainment experience.

That's entertainment...2015 style

Although most TV viewing will still be to scheduled programmes via the TV screen by 2015, the way we access, consume and share our TV viewing will feel quite different.

The main changes that are likely to emerge by then include;

- Better EPGs, to improve navigation, enabling people to get even more of the TV they love. There may well be an EPG battle between platforms and connected set manufacturers, who will install proprietary EPGs. Either way, smarter EPGs can lead the viewer to even more engaging content, while companion devices can act as real-time EPGs, using apps like zeebox.

* Synchronised content, from both the broadcasters and third parties, including advertisers. I believe there is a strong opportunity for advertisers funding content on second screens to enhance the content on the main screen. Such content is likely to increase appointments-to-view, consolidate audiences, increase live viewing and enable more effective advertising response.

* Personalised timeshift. It is highly likely that a great deal of timeshift viewing in future will take place when people are 'on the move' as mobile devices become a bigger part of the TV ecosystem. Viewers will have the opportunity to pick and mix the parts of a programme they want to record – selected highlights from a sporting event, specific items on a news report or a five minute resume of last night's soap episode.

* TV on all screens...and all screens linking to the TV set. This does not just mean that a programme can be viewed across all screen-based devices in the home, but that the content on all other screens – including synchronised content – can be linked directly to the TV screen.

* More intuitive sharing of content – through recommendation, real-time shared timeshift viewing and increasingly integrated methods of merging social media with the TV viewing experience.

* Deeper interactivity. The interactive involvement with a programme can be integrated, deepened, lengthened (e.g. so interactivity can be encouraged between programme

episodes)

- More content to access. Premium movies, obscure sporting events, US series premieres, non-broadcast content and just about any archive content in existence will be available direct to the TV set.

- Greater theatricality – via bigger screens, better sound, 3D or Ultra HD. Consumers value theatricality more than for any other aspect of TV, which should help to maintain the primacy of broadcast TV and the TV screen.

All of these trends should make TV viewing more appealing and its impact on the viewer more powerful than ever. A pen picture of what it might mean for the average household could be as follows;

Meet the Brennans. They are fairly average family of four from 2014 (or an early adopter family from 2011, as most of these TV-based activities are already potentially available to anybody with the right technology and a slightly above average understanding of how it works). Like almost every other household in the land, their main family room is still furnished around two fixtures; the fireplace and the TV set. Both provide central heating of a kind, either physical or emotional respectively. There are still things known as TV sets in the kitchen, the master bedroom and one of the kid's bedrooms (the other uses their laptop as their main entertainment device and streams or downloads TV content through that). There are also a number of boxes attached to these sets; games consoles (which can also play content from most on demand sites), a Sky HD box on the main set and Apple TV attached to the master bedroom set that can also stream to other sets in the home.

Supporting devices are also evident. David has an iPad, which is normally at rest on the coffee table, while his wife, Jules, rarely watches TV without her Samsung Galaxy smartphone clutched in her hand or close by. Eldest child Nicole, 19, owns a laptop, which she mainly uses for student coursework, as well as entertainment (often at the same time), while 12 year old Sean only has eyes for his Nintendo DS or the PS3 attached to his bedroom TV set.

Jules comes home and decides to watch a bit of TV before starting on the evening meal. She checks via her smartphone app to see what is available; not only what is on the Sky EPG but also notifications about new programmes that have been recorded onto her (personalised) planner, and an Anytime+ pick of the night selection based on her reported preferences, recent viewing behaviour and friends' recommendations. Almost paralysed by choice, she receives a notification from best friend Ally, who is about to start watching 'Next Top Model...' (timeshifted from the previous evening) and Jules agrees to watch it with her. One click on the notification and the programme is on TV. Jules and Ally then spend half an hour commenting on the clothes and the poses via mobile whilst watching the show through to its conclusion.

David comes home and immediately requisitions the main set for the live football on Sky; Jules transfers the viewing of NTT seamlessly to the kitchen set, where she resumes her ongoing commentary with Ally. Nicole walks by, spots an outfit that one of the models is wearing, and waves her own mobile phone at the screen. Immediately, the details of what all the models are wearing on this week's episode comes up on her smartphone screen, together with stockists and pricing. She orders a size 8.

Meanwhile, David is gearing up for the kick off. It's Manchester United playing Arsenal and he sets up the 'chatter bar' at the bottom of the screen, deciding whose comments are worthy of inclusion in his own personal 'fan zone'. It is the usual group of United-supporting friends, but he has also added in a few Arsenal supporter mates so he has somebody to ridicule, should United win again. Bruce from NZ, Paul from Australia and Tommo from Watford will all be virtually sitting in his front room with him while the match is on and he's looking forward to the banter, which he controls from his tablet but is also played out on his TV screen. They also all play the in-match betting game for points only. David will occasionally bet real money on his picks, but not for a game this important!

Sean enters the room when United go a goal up. He loves football, but only if his team are winning, and he immediately checks his Fantasy League on dad's tablet to see how many points Rooney's goal has added to his weekly score. He sits to watch with dad, simultaneously checking on his ongoing Nintendo DS game he's playing online with his mate from school. He has to break off, however, at 8.30 because 'Test The Nation' is on Channel 4 and he is top scorer in his school league and heading for one of the prizes on offer from the programme sponsors. He has to watch live, because it is the only way you can play along and, besides, the scores are compiled online almost immediately, and he'd already know both the answers and the results if he watched in timeshift mode.

After the football, the whole family is gathered to watch a bit of TV together. They have decided, after a great deal of discussion and compromise, to watch a first run movie as a treat. They choose Ryan Gosling's latest comedy from Sky Blockbuster, but after 15 minutes the kids decide they want to watch the rest of the movie in their bedrooms, streaming it to their nominated devices. David manages another half an hour before calling it a day. He grabs the rest of the movie on his iPad - there is just long enough left to fill in the train journey in the morning. He also records his personalised Match of the Day highlights programme, featuring just the games in which he has an interest, ready to download directly to any screen he selects.

Jules settles down for some 'me-time' before she too calls it a day. Rather than graze through the schedules or plough through the on demand options, she gets her 'TVGenius' to provide her with a specially-prepared, personalised collage of what happened in her two favourite drama serials and the results announcement from 'Next Top Model'. She is also asked if she would like to see a trailer for a major new comedy show from Channel 4. She's intrigued enough to watch a short clip, laughing out loud as she sets it on her planner. She has the house to herself tomorrow night, so she adds it to her own 'Jules TV' schedule, set around the live broadcasts of 'The X-Factor' and 'Ant & Dec's Celebrity Schmooze' (because she wants to vote live for one and play along with the other). Sorted[62].

This pen picture is not exactly like real-life yet; I'd never be allowed to watch football if my wife was cooking dinner! Still, many of the activities and choices mentioned are already with us in some shape or form, and availability increases every time somebody buys a new TV set, tablet or smartphone. The networks and content producers are already enabling the sharing, distribution and enhancement of their most popular programmes, linking content and screens via wi-fi and portable smart devices. There is a relatively short way to go in terms of functionality to make this type of viewing possible - it is more about education and desire. The audience needs to understand what is available and be able to manage it easily and intuitively. Fortunately, television is in a great position to promote such activities, as we have seen repeatedly over the decades.

ADVERTISING

Reading the pen portrait of how we may view television within the next few years, you may be forgiven for thinking *'that's great as far as TV programming is concerned, but it doesn't leave much room for advertising'*, but that is not the case. For one thing, through sponsorship, advertising-funded programming, product placement and other forms of editorial marketing, brands will be featuring within and around the editorial as a matter of course. They will often be so deeply integrated, that it would be impossible to highlight their presence through a pen portrait; much of their impact will be at an implicit level.

There will be numerous opportunities for brands to be an integral part of that viewing experience, and some unique opportunities to generate response or interaction with advertiser content. Perhaps the biggest change will be advertisers thinking beyond *'should I go sponsorship or spot?'* and starting to use the full array of marketing tools that television is beginning to provide. Perhaps the advertisers can enter the fray in the following ways;

When Jules switches the TV set on, she is drawn to an ad on the EPG for the new boxed set of the twelfth series of 'Shameless' and decides to order it. She uses her Sky Points, awarded for previous transactions, to get a significant discount. The series is automatically downloaded to her planner, but she also receives the physical DVDs in the post as part of the deal.

The EPG also contains ads for the latest blockbuster release from Pixar and she knows Sean would love to see it, so she downloads the trailer onto the planner to show him later. The planner has different files for specific content; for example they have a 'New Car' file which contains all of the latest TV ads for cars within the family's price range; they will sit and watch together before coming to a 'family decision' (although Jules has already made her mind up what the family decision should be!).

As she watches 'Next Top Model', she can't help but love the handbag featured in one of the fashion shoots. She calls it up on her Smartphone's 'NTT' app, which immediately provides product details, pricing and nearby store locations. She decides not to buy off screen this time round, but does log the details of the nearest store for when she is in town next.

Jules also likes to play along with the commercial breaks. Sky is running a 'Break Bingo' promotion (sponsored by Foxy Bingo), and every time she sees one of the featured ads on her card, if she presses red on the remote, she marks it off her card. Only two more and she gets a £20 voucher towards more downloadable content from Sky Blockbuster or Sky Sports.

As Jules swaps rooms to watch 'NTT' on the kitchen TV set, her friend Chrissy sends her a message to say she MUST see the new M&S food ad on C4 – it has a brilliant recipe for a spicy chorizo and prawn *linguine. She would have missed the ad itself on the journey between rooms, but Chrissy's link means she can watch it 'as live' before going back to where she left off on 'NTT' or she can save it in her 'Recipes' file on the planner. She decides on the latter.*

David is too engrossed in the football to interact very much with anything; in fact, he has switched all other phone and messaging systems off so he can concentrate on the game. During the half-time break, though, he does see a couple of ads he likes and passes them on. There is also a pre and post break football quiz, sponsored by Carling, which he plays along with. Who knows, he might win a year's supply of lager...

Sean is working hard to win some new Nintendo games. He is particularly keen to get Modern Warfare 8 – Invasion of New York. There is an app for The Gaming Channel which allows him to download segments of the trial content from the on-screen ads straight to the console, and if he can complete the first mission quickly enough he can get a major discount on the whole game - as well as the status of appearing in the top performer's league positions displayed in the channel's interactive zone.

David is still waiting for the second half to start. He absent-mindedly presses red to order free samples of the new Cadbury's chocolate bar, then collects another Meerkat on his mobile from the TV ads (which feature all of the new CompareTheMeerkat characters, which will tell jokes for you once you have collected with a simple wave of his smartphone). He sees an ad for a new flavour of his favourite soft drink, Vimto, so he presses the Tesco app on his iPad and automatically downloads the product into the family shopping basket.

If he'd asked Jules to get it she would have ended up ordering Ribena, and that would be no good at all!

Within that pen picture, we can see a host of ways in which advertisers can get closer to their consumers via a more engaging, social and interactive TV viewing experience. There will still be room for spot advertising – indeed for some years to come it will continue to be the main revenue generator – but there are many other ways to extend, enhance or replace it with new ways to communicate with and appeal to the audience.

In this chapter, we're going to investigate where advertising (which I'm using as shorthand for brand communications in general) is going and TV's future role in the media mix. This will just be taking in the current trends we're seeing in TV (longer-term trends will be examined in the following chapter). We can explore it through the following three paradigms;

- Brand-building vs response
- Spot vs. content-related
- Beyond broadcast

Brand-building vs. response

Traditionally, TV advertising has always been either brand-building or response. Today, most TV commercials are both - the brand-building properties of engagement and emotion are well-recognised, while TV's ability to generate response from anywhere is reflected in the fact that the vast majority of ads (around two out of every three) now have some form of call to action included, usually in the form of a URL.

As the pen picture demonstrates, the boundaries between brand-building and response advertising will continue to blur and most TV campaigns will have a response element built in. The range of response mechanisms will grow and brand activation will be about much more than website visits or product purchase, integrating with social media activities, driving viewers towards communities of interest or creating multi-layered responses across many touchpoints and over long periods of time.

How we evaluate TV campaigns across both brand-building and response will be important. Existing metrics, such as brand tracking, have been proven to be only faintly related to actual product performance, so we need to unlock the emotional and implicit measurements of performance to fully understand TV's contribution to the brand's bottom line. At the same time, we will need to integrate response metrics, such as analytics or econometrics, so that the return on media investment is accurately calculated across both the short and longer terms. If that can be achieved, then television can only become even more central to the role of advertising in the 21st century.

Spot vs. content-related

During the great TiVo scare, advertisers rushed to find ways to become more integrated into the surrounding content in order to counter the looming threat of ad avoidance. In fact, it proved unnecessary; spot advertising still accounts for well over 90% of TV advertising revenues, but in the process the advantages of content-related advertising (sponsorship, advertiser-funded programming, product placement etc.) have become much better understood, leading to the conclusion that these forms of TV advertising will grow faster than the spot market for the foreseeable future.

Sponsorship – A brand's best friend?

Although brand sponsorship of programmes is part of television's history (after all, it is the reason we watch 'soap' operas), in the UK and parts of Europe, regulatory restrictions have meant it is a relatively recent way for brands to reach TV audiences, it is only in the last decade or so that they have evolved from being little more than billboards at the start of a show to becoming much more creatively integrated.

It is fair to say, though, that we are only beginning to understand how sponsorship works differently to spot advertising.

In 2007, I was asked to present an overview of what we knew about how TV programme sponsorship worked. After several weeks of trawling the academic and trade publications, I came to the conclusion that we know in fine detail how well sponsorship works (every broadcaster regularly conducts case history material on all major sponsorship campaigns) but there was precious little about why that should be, and how sponsorship performed differently to spot advertising. When I presented this, I was challenged to put it right!

A year later, we had finished our major study – "*Sponsorship – A Brand's Best Friend*[63]' – which went on to win the IPA's Simon Broadbent Award for best media research paper at the 2008 MRG Conference. It used a mixture of methodologies – surveys, focus groups, ethnography, lab tests and implicit attitude testing – to demonstrate that it was the power of the association, rather than the sponsorship content itself, that generated the payback. If the creative could integrate between the programme and the brand, while also framing the show in a way that introduced the brand into the viewing experience, then a 10 second sponsorship bumper could convey far more about a brand, implicitly and explicitly, than a spot ad several times that length.

We tested the performance of a range of current sponsorship activities, and were able to show how sponsorship worked in different ways. For example, a strand sponsorship (such as Toyota's sponsorship of Channel 4 drama) was really good at gaining stand-out and a sense of scale (especially for smaller brands) while programme sponsorships were better at creating emotional associations with the brand. Implicit attitude testing also demonstrated that many of these effects are below the radar, that our real feelings and perceptions about a brand are impacted in ways that we cannot bring to our conscious minds.

But perhaps the most powerful effect of all was sponsorship's ability to create a perception that a brand is 'big' and 'important'; to make it famous. After all, reasoned one respondent after another, you have to be a big brand to be able to afford a sponsorship (because their perceptions are that it is relatively expensive and the brand needs to commit to the whole series) and you have to be confident about your future and who you are. This sort of perception works just as strongly in advertising as it does in the mating rituals of the animal kingdom (think the peacock's tail again) and goes a long way to account for why big TV advertisers consistently outperform their rivals. It is all about the mating game.

Just as in the mating game, for every winner there must be a loser or two. Neuroscientists have found that, when viewers are exposed to sponsorships for a brand, this often also has a detrimental effect on competing brands. As Peter Pynta, Neuro-Insight's Marketing Director wrote recently;

> *"What we found fascinating was the effect of sponsorship and product placement on the competitor brands that were not featured in the program. We found that if a brand increases its strength as a result of sponsorship or product placement, the competitor's brand strength was weakened. Not only do brands compete in the marketplace but they also compete in the consumer's brain".*

Sponsorship creates fame and a sense of a brand being bigger than it is, and it communicates deeply through what the association with the programme says about the brand. Most of the hundreds of case studies also show it works in driving the usual metrics, such as awareness, brand preference, online traffic or retail footfall. It also suppresses competitive brands in our minds. Sponsorship works, which is why UK advertisers now spend around £200 million a year on it. However, sponsorship is still undervalued, which is why it isn't twice that figure.

When sponsorship is not enough – advertiser-funded programming

Advertiser-funded programming (AFP) is now allowed in the UK, bringing it into line with most TV markets. At its best, AFP allows advertisers to create the perfect sponsorship vehicle for their product, where none naturally occur in the schedules, while for the broadcaster it brings in additional money and/or funding for the programme budget.

In some countries and on some channels, AFP is the source for whole chunks of the schedule...and it shows. It always seems very clear when AFP is being used to provide cheap, off peak programming rather than as a finely-crafted way to support and promote brand values through high production values.

AFP is also resource-hungry, bringing broadcasters, advertisers, agencies, production companies and possibly AFP specialists into sensitive and time-consuming negotiations. No wonder so many fall at the first hurdle and AFP revenues have failed to match early projections. The high relative cost compared to sponsorship, especially when the brand cannot feature on the programme and viewers still see it as a sponsorship, have also contributed to its disappointing performance. Then there is the common problem that, often, viewers are not as interested in an advertiser's preferred subjects as had been thought. So, we have seen many AFPs struggling to reach a viable audience (the off-peak slots for many of them have undoubtedly contributed to that) and/or fall foul of the strict regulatory environment.

Recently, though, as brands find ways through the minefield (helped by a growing number of specialist companies) and understand how AFP can best work for them, the quality and quantity of advertiser funded programming is growing. In the UK, the government agency COI (Central Office of Information) commissioned several successful public awareness series, such as *'Beat: Life on the Street'* (about community policing) and *'Celebrity Quitters'* (with the NHS, aiming to help people to quit smoking) because they have genuinely interesting content and a relevant agenda. More recently, we have seen more and more peak time AFPs, such as *'The Krypton Factor'* (funded by Sage Business Software) and *'Britain's Best Brain'* (funded by Nintendo DS).

The work involved in bringing a high quality AFP to the screens means it will never be as widespread as sponsorship, but it has passed a very long teething process and is finally starting to bite.

Product placement – deep-diving into the programme itself

Creating a clear boundary between advertising and editorial has been a guiding principle for regulators everywhere, especially concerning television. The concern that TV's unique ability to influence and 'brainwash' us needed to be tempered with rules and regulation has particularly related to product placement, where the product is embedded inside the programme content. Even when the EU opened up the market for paid-for product placement, as part of its AVMS Directive, the UK held out for almost two more years before finally allowing it in early 2011.

Any doubts have so far been unjustified. This is partly because UK viewers were already experiencing product placements within US programming, where series such as *'Ugly Betty'* and *'30 Rock'* did not just feature products but created whole scenes around them. The US has already taken product placement to its extreme position, and UK viewers either didn't notice or didn't mind.

Nielsen has been monitoring audience reaction to product placement in the US and the UK for several years. They have conducted hundreds of thousands of interviews, and concluded that viewers are quite positive about the concept of product placement, many even welcome it. They would reject 'fake' placement (such as James Bond driving a Ford Mondeo) but enjoy the fact that real brands add an air of reality to their favourite shows. For example, long-running soap *'Coronation Street'* has always portrayed regulars at the Rovers Return pub drinking fictitious Newton & Ridley beer, which has driven many fans to call for a more realistic depiction of their alcoholic intake.

The Nielsen research also shows that, when they are aware of a placement, viewers tend to have a much higher opinion of the brand in question. But this is just the tip of the iceberg because the best placements are not meant to be noticed. However, work done by neuroscientists has demonstrated that the halo effect on brand engagement can be measured even when the brand is only on screen for a second and totally below the consumer's level of awareness[64]. Similarly, ITV eye-tracking research demonstrates conclusively that our eyes are drawn to brands depicted in TV programmes, even when they are in the background.

Product placement is still finding its feet in the UK, although ITV believe it can be a £50 million market within the next couple of years. Its cause has certainly been helped by the ability of producers to digitally insert the brand into the scenes rather than having to tie the deal up at the pre-production stage. As the ultimate example of brand integration into programming, with all of the associational benefits that brings, most experts expect to see product placement revenues grow exponentially over the next two to three years.

Getting closer to content in an online world

In recent years, we have seen the emergence of new commercial models to bring brands increasingly closer to the content we love and share, and much of this activity occurs in the online space. Whether it is advertorial, branded entertainment, AFP, sponsorship or product placement, brands are using online to enhance the associations, generate brand transactions and move viewers closer to purchase. Amongst recent examples, British Gas offered an interactive game around their AFP *('Green Up Your Life')* successfully encouraging kids to become more aware of how they could reduce their carbon footprint. Furniture retailer Harvey's, who took over sponsorship of *'Coronation Street'* from long-term predecessor Cadbury's, ran an interactive area within the show's red button site that attracted more than half a million visitors a month and redemption rates peaking at 50% (many times above the industry average). Chocolate brand Bounty, as part of its sponsorship of *Love Island'* created an interactive area that attracted over 1.2 million people in one month alone, and a dwell time of close to 10 minutes. Sainsbury's got close to the *'X-Factor'* by sponsoring the food section and catering for the housemates, that was featured weekly on the programme's website – seen by over 650,000 viewers for an average of 12 minutes each and driving a significant number to Sainsbury's own website.

Online allows advertisers to get closer to content but then move to front of screen once viewers move from TV to companion screens, enhancing the associations but also successfully encouraging participation, transaction and even purchase.

The Advertising Break

The growth of content-related opportunities should not detract attention from the humble 30 second TV spot. The resilience of the commercial break to retain viewers in the face of ferocious competition has renewed advertisers' confidence in TV, which is beginning to be reflected in levels of creativity, production budgets and media spend around the spot. 'The 30 second spot' is a lazy generalisation anyway, as spot strategies can be far more effective, employing blipverts, top and tails, roadblocks, break-overs and interactive spots, to name but a few.

It is the creativity of the advertising itself, though, that will become increasingly important, and advertisers are already starting to recognise this. The power of a highly creative commercial to engage and influence has already been well documented, but special mention should go to the recent IPA study into creative effectiveness[65]. Analysing over 250 recent IPA effectiveness award winners, it found that creatively-awarded TV campaigns were <u>12 times</u> more successful in shifting market share per £1 spent (compared to 3 times more successful just five years ago – a sign that creativity is becoming more important at separating the winners from the losers, not less. Creativity sells, for all of the reasons I have outlined, which is just as well - because the viewer expects it.

Generally, the new generation of connected TV viewers is as sanguine about the ad break as anybody else; it is a necessary evil but otherwise doesn't impinge on their consciousness very much at all. There is little evidence that the ad break is about to be edited out of their lives, but neither is there a sense that it is transforming before their eyes; the average ad break remains similar to what they have been used to for decades. Maybe that is its strength.

Broadcasters are introducing their own initiatives to make commercial breaks work harder. Channel 4 has introduced themed breaks, around the theme of the surrounding programming. For example, comedy themed breaks around programmes like '*The Great Ormond Street Comedy Gala*' (where top comedian Jimmy Carr gently spoofs the ads), breaks themed around topics such as 'Great British Design' and movie promotions have all been successfully introduced. Not only do such initiatives keep viewers more engaged with the advertising, but they significantly improve the ability of the break to retain those viewers.

There is, though, one area of the commercial break where those who have experienced TV's near future expect to see change

- the number and relevance of TV ads that they are expected to sit through. They sense that the next wave of TV technology will mean that the less relevant ads are somehow edited out. They also expect to see fewer ads in total (which, to be fair, is to be expected considering they are sitting through an average 10 ads per day more than they were just a decade ago). Broadcasters will need to walk a fine line if they are to ensure ad breaks make money without shedding viewers.

Beyond broadcast – what ad formats do viewers want?

The beauty about taking people to the future and then asking them about it is that they can answer questions about different creative formats and interactivity, with a strong sense of the context within which these new forms of advertising will be viewed.

The Tellyporting research tested a range of different formats which are already being used within an on demand environment and which could soon be a regular feature of broadcast ad breaks as well. We were able to test concepts such as interactive sponsorship bumpers, in-skin advertising around short-form video, 'choose your own ad' formats (allowing viewers to select from a range of ads), digital insert (similar to product placement) and standard TV ads with interactive sidebars or overlays.

The encouraging finding was that viewers generally have a very clear idea of how each format could work and how it affects the commercial contract. So, in-skin would only work for short-form programming such as music clips; interactive sponsorship bumpers should include an element of the show being sponsored; and digital inserts should work by being so integrated into the programme that they are not even noticeable as 'advertising'. Choose your own ad was welcomed by many as the first step in producing more relevant and engaging advertising experiences, although a minority didn't want to have to undergo yet another selection screen in order to get to their on-demand content. However, perhaps the most encouraging reaction was to TV ads with interactive side bars, a format that has already been successfully introduced into on demand shows in the UK by ITV.

Although this format is just a variation on traditional spot advertising, with the option of interacting via an interactive sidebar, its sheer transparency and ease of use saw viewers perceive this as the 'advertising of the future'. They appreciated the fact that they could turn the side bar on or off, that responses were prompted and easy to access and that it could take them on a 'journey' with relatively little effort on their part. In fact, the less the effort they had to make, the greater the chances of persuading them to interact with the ad at all. They also liked the fact that they could respond without having to leave the programme they were watching (a common complaint against red button advertising).

Once Tellyported families understood and experienced these new forms of advertising, they universally accepted the possibility of engaging with them. They were full of ideas that would encourage them to respond and interact. They suggested a 'red button' that would allow them to capture favourite ads and store them together on their planners. They identified a range of calls to action that would appeal to them. They could all think of ways in which they would respond, link to social media, bookmark or even purchase, just as long as it was easy, quick and didn't take them away from what they were originally watching.

Tellyporters didn't want these formats to be intrusive, but they wanted them to be intuitive and part of a narrative (they frequently spoke of being taken on a journey). If advertisers get this opportunity right, the viewers will take part. They are already primed.

To finish this chapter off, I'm going to list five ways in which TV advertising <u>will </u>change over the next five years, all of which offer massive opportunities for advertisers.

1. **Interactive everything**: currently the majority of TV ads feature a call to action, so the idea that most of them will have a direct interactive or transactional proposition is already well-established. This will go beyond spots to also feature interactive sponsorship bumpers and links for brands using product placement.

2. **Break gamification:** it will be vital to keep viewers within commercial breaks, and 'playfulness' is a major part of the TV viewing mindset. Some broadcasters are addressing this issue via themed breaks or competitions but, where regulation allows, we can expect to see more opportunities for the audience to play along, often sponsored by advertisers.

3. **Screen jumps**. Companies from car manufacturers to retailers are already experimenting with ways to shift transactions from the TV screen to other screens. This simple, intuitive approach to moving content and services across screens and platforms offers a truly converged and integrated experience for the viewer which should significantly improve response rates and customer conversion levels. ITV's recent exclusive deal with Shazam shows that the TV set does not even need to be connected for such synchronization to take place.

4. **Greater content integration.** Although the spot is not going away, we can expect to see more novel and effective ways of integrating brands into programme content... and then enhancing that relationship online, through programme and broadcaster assets.

5. **Greater integration with social media** from recommendation-based programme and ad viewing suggestions to automatically linking social media accounts to TV viewing all the way through to integrating social media content into the broadcast stream.

YOUTH

If this section is to address all of the issues relating to TV's connected future, it must end with an investigation of the youth audience. Whenever TV's obituary writers run out of evidence for the inevitability of its death, they usually resort to the 'digital youth' argument - young people today have grown up with digital and are using it in ways we cannot imagine, so no matter how we poor analogue dropouts continue to use media, <u>they</u> will be different!

The youth of today. We've been treating youth as a separate species since at least the 1950's, when the baby boomers finally began to hit their difficult teenage years. Ever since, we've worried about the things that have defined them; the clothes they wear, the music they listen to and the drugs they ingest. 'Youth' is a strange and threatening country. We are now told what distinguishes youth isn't their tribal affiliations through music, fashion and culture but their ubiquitous and intuitive use of digital technology, which signals the end for traditional media.

It is undoubtedly true that the youth of today are watching less television than the older segments of society, but then they always have. Ever since ratings measurement began, teenagers have watched on average between 30% and 40% less television than the adult average. That hasn't changed. After all, they still date, get drunk, sit up all night listening to music and lie in bed for far too long in the mornings to possibly fit the same amount of viewing time into their already busy diaries. *Plus ca change...*

During 2011, young UK adults (16-24 year olds) watched 2.75 hours per day of television, slightly down on the 2.81 hours recorded in 2010 the highest recorded level since 2001, but still well up on recent years[66]. They viewed around a third less than the all adult average of 4.03 hours per day, just as they have for decades; hardly a crisis and certainly not a 'migration'. They are behaving as they always have.

If we look just at the teenage audience, there is some evidence of changing behaviour. There has been a noticeable drop in teenagers' live TV viewing. All of the new opportunities to view online, together with the sheer variety in the ways they can nowadays fill their time, has meant that today's teenagers especially the 16-19 year olds – only watch about half the amount of broadcast TV that their older counterparts do.

The big question is; is this structural or cyclical? Does today's crop of teenagers exhibit such wildly different behaviour from their predecessors that it will irrevocably affect their future viewing habits? Or is their behaviour specific to their lifestage? What will happen when they reach adulthood, start working for a living (if they are lucky) and come home at the end of the day tired out and in need of a good chill-out in front of the box?

The BBC answered this question in 2010, using BARB data to track young people's viewing behaviour as they passed through the lifestages. One of the advantages of a continuous ratings panel is that the same people often stay on the panel for a number of years, so the BBC's analysis showed how it changed.

Three 1-year cohorts of people on the panel; those born in 1985, 1990 and 1995, had their total TV viewing tracked for each year they grew in age. As we can see from the graph below, each successive cohort has viewed significantly less TV during those wild, experimental, sociable years between the ages of 16 and 20. Their viewing via the non-BARB platforms, particularly web TV, would make the gap less if it was included, but even so there appears to be a drop in TV's appeal for each successive cohort of teenagers.

What is interesting is what happens to each cohort when they reach their early twenties, and join the adult world. For the two previous cohorts – those born in 1985 and 1990 – their viewing immediately hits similar levels to their parents and older siblings Although we only have one year's worth of data for the 1990 cohort hitting 20, it is startling how quickly they slip into the viewing habits of the baby boomers. They almost doubled their hours of TV viewing in just a couple of years.

The question still arises - will the latest cohort, born in 1995, exhibit the same behavioural characteristics and return to TV as their first choice for entertainment? They are watching two hours a week less broadcast TV than the previous cohort, but the evidence suggests that, from aged 18 onwards, TV viewing will rise sharply.

TV Viewing Hours by Age – Cohorts Born in 1985, 1990 and 1995

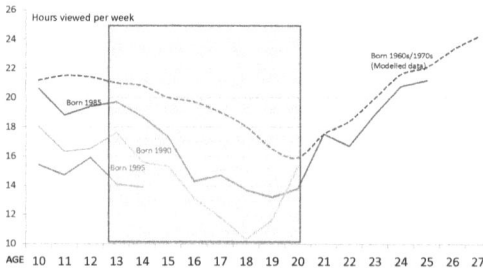

Source: Broadcasters Audience Research Board (BARB). Reproduced by kind permission of the BBC

Figure 3

While we're on the subject of young people's viewing levels to TV, let's pause to reflect on the age when viewing drops to its lowest point - around their eighteenth birthday. This is obviously an age when they have more freedom to go out, which they exercise with great relish. Part of that freedom is based on financial (and sometimes physical) independence, via their first jobs, a gap year or full-time education. If it's the latter, many of them will be living away from home, which suddenly means they drop off the viewing radar as ratings panels don't generally cover them. Estimates of the number of young people that this potentially takes out of the UK's viewing universe range from 1m – 1.5m, which is a sizeable chunk of the late teens audience. The big question is - how much viewing are we missing?

Students used to be notorious for watching TV indiscriminately. From *'Countdown'* to *'Deal or No Deal'*, the popular image of students lying in bed watching daytime TV lives on. Apparently, it is not wide of the mark. A study I commissioned three years ago, 'The Secret Life of Students[67]', documented the popularity of *'Bill Oddie's Springwatch'*, *''Loose Women'* and even *'Countdown'* itself, while communal viewing of event TV or cult favourites (such as *'Peep Show'*) meant that often whole evenings in were planned around the television schedule.

The possibility that students living away from home watch even more TV than the rest of us is made clear by data from Freewire, who pipe IPTV services into around 150,000 rooms in halls of residence, and A C Nielsen, who started to monitor campus students in the US just four years ago. Both studies calculated that full-time students watch three and a half hours of broadcast TV a day on average; that is a lot of viewing to take out of the ratings equation.

It ain't what they view, it's the way that they view it

Although the television industry will still need to work hard to remain relevant to the digital natives, there is cause for optimism that the current generations have not fallen out of love with TV. But, yet again, the apparent stability of the headline figures disguises some huge changes in how they view the medium.

First of all, much more of their viewing is on demand. The linear schedule is still responsible for 90% of all TV viewing, but for children and youth audiences, the figure is closer to 80%. On demand accounting for approximately 5% of their viewing, compared to 2% for the general population[68].

Secondly, far more of their viewing is accompanied by other screen activities. Touchpoints records more than a third of the time 16-24 year olds watch TV they are also online. As we have seen in previous chapters, this should be viewed as more of an opportunity than a threat to TV, and their 2-screen behaviour is already providing some glimpses into the future.

Much of that concurrent activity is around social media. The latest Touchpoints shows that, for 16-19 year olds, social media accounted for almost half of all their time spent online – even more when they are also watching TV - and is proving to be increasingly influential in many aspects of their everyday lives (including the TV shows they choose to watch).

As with all other demographics, they appear to have their own version of me-time, we-time and in between-time, but more of it is <u>we-time</u>. The fun of sharing viewing experiences with friends still figures highly, and many of them still use certain programme events as a reason to hang out with their friends or family. Consequently, young viewers watch a greater proportion of their TV with other people, according to BARB data (although viewing via other screens appears to be a more solitary activity).

Finally, they have their different channels and programme preferences, although their top programmes still overlap a great deal with older age groups. The top ten programmes for 16-24s regularly include *'The X Factor'*, *'Coronation Street'* and *'Eastenders'*. One of the things overlooked in the race to dismiss TV's role in young people's lives is just how much television content is specifically made for this age group these days. We now have programmes and whole channels that can appeal to young audiences more efficiently than many specialist magazines.

Young people ultimately keep watching TV, though, not because of a steady influx of new MTV sub-channels, but because TV fulfils particular needs states, most of which had not even been considered when television's obituaries were being written.

Generation Whatever

I've been involved in numerous studies looking at youth and their media use, particularly television. The things that have maintained TV's relevance to their lives have been surprisingly consistent over the decades, but expressed in superficially different ways.

The most insightful study was labelled 'Generation Whatever[69]', mainly because that was young peoples' approach to technology in general; it is not embraced as the next big hope, it is mainly used with a 'whatever' shrug and a scattergun focus.

The study identified and explored the key needs states that TV satisfied for the 'youth' (which can range from 8 year old kids to 21 year old full-time workers). It was based on more than 1,000 online interviews, hours of buddy-pair interviews and an open-ended ethnographic observation of their lives, and highlighted the four core needs that TV satisfies across Generation Whatever;

Emotional central heating - as described by the respondent I quoted in Chapter 1 - refers to the comfort, familiarity or warmth that comes from just having the TV set on. TV was often described as good company when alone, the sense of having another human being (or several) in the room. In emotional terms, it's a radiator, not a drain. This may mean television has a back seat role at times, as the digital natives text their friends, update their status and do their homework (often simultaneously), but it is a role that any other media channel would find very hard to replicate.

Glamour and entertainment might seem superficial, as needs states go, but the Generation Whatever research re-emphasised the importance of entertainment in their lives, linked to a desire to experience or have a window to a glamourised version of real life. It brings in the importance of celebrity, aspiration, identity and cultural learning. It also reflects the frantic need of many teenagers to fill in dead time and to stave off boredom, preferably without requiring too much effort.

Social currency has already been discussed at length elsewhere, but is of even greater importance to Generation Whatever, especially from the early teens onwards. Television not only fuels their conversation, but also provides opportunities for badging and reflecting their developing sense of identity, via favourite programmes, celebrities, genres and characters. From their perspective, you don't want to be left out of the playground or bar conversations because you haven't seen '*X Factor*' or '*The In-Betweeners*'.

Status and aspiration is particularly important at a time of their lives when young people are working out what is 'real' and what is their place in the world. Television provides a window on the world, in the same way it is often used by recent immigrants to better understand their new host culture and language. It provides an understanding of how the world works, what is acceptable and, most importantly of all, what their place in it could be. One particularly good example is career aspirations; there are countless examples of TV programmes that have influenced young people to take up a career in a diverse range of occupations that include forensics, veterinarian surgery, community policing and modelling. Recently, Channel 4's *'One Born Every Minute'* was cited as the reason for an increase in young people applying for a career in midwifery.

Obviously, television is by no means the only media channel in young people's lives, and it is under much more sustained competition for their time and attention than ever before. They watch less television than average and more of the TV they do watch is juggled with additional activities. That said, the Generation Whatever research reported that TV was still the first place they turned to when bored, or hanging out with friends or family, or simply filling in time. They still talk about it more than any other media activity (including online) and they still claim to value it. Where this can be seen in its starkest form is through their attitude to and knowledge of TV advertising.

TV advertising - less exposure, greater impact

In all my years researching media audiences, one consistent theme has been the ability of young people to recall and engage with far more television advertising than older adults, despite watching so much less of it in volume terms.

There has always been a theory that lighter viewers are impacted more by the TV ads that they do see, because there is less 'noise', so the good ads stand out. This principle could certainly be applied to the teenage audience, but their undoubted enthusiasm for creative TV advertising, and their superior knowledge of the brands being advertised, probably owes an equal amount to their enthusiasm for brands and an almost post-modern appreciation of the creative process.

Most surveys will show 16-24s having higher recall levels for a wide range of brands and their advertising, not just those that are specifically aimed at them. The recent Deloitte survey for the 2011 Edinburgh International TV Festival[70] showed TV as having the highest impact and memorability of any media channel, by quite some distance. However, the distance between TV and the rest is even more pronounced among the 18-24 year old audience, as the following two charts demonstrate.

Figures 4 & 5

Advertising impact by media channel

■ All adults ▨ 18-24 adults

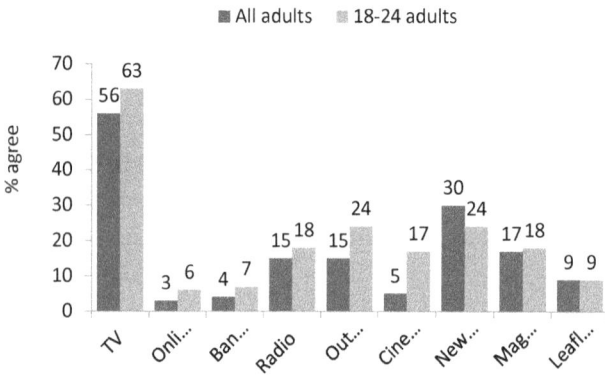

Source: Deloitte/IPSOS August 2011
Base: Responses to question "which of the following types of advertising do you think have the greatest impact on you personally?"

Source of most memorable advertising campaign

■ All adults   ■ 18-24 adults

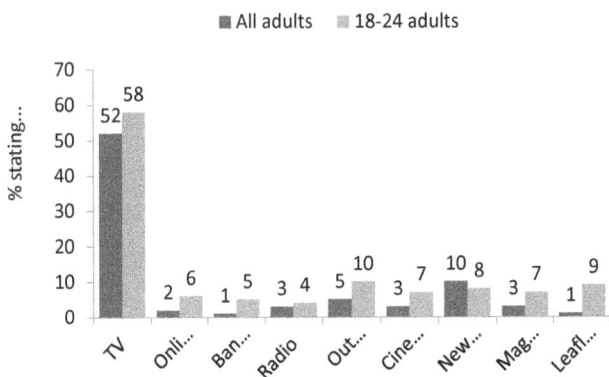

Source: Deloitte/IPSOS August 2011
Base: Responses to questions "which ad campaign do you consider to have been most memorable since the start of 2010?" and "on which media channel did you encounter it?"

Reproduced with kind permission of Deloitte

The youth audience is more likely to say TV advertising has the greatest impact on them, or to name a television ad (almost exclusively) when they are asked to think of the most memorable ad campaign of the previous year.

This is a phenomenon which is consistent over time. Twice, in the past three years, I have been involved in projects which required us to collect interviews with teens on film. Each time, the camera crew stationed itself at the top end of Carnaby Street in London and approached every trendy looking teen they could find, to ask them one simple question; *"what is your favourite advertisement"?* Each time, the teenagers responded at length, often acting out the ad itself, or replaying the script word for word. They knew the brands inside out and would often volunteer their rationale for the creative and what it said about the brand. The interesting thing was that, overall, we collected answers mentioning more than 80 brands and, even though we only asked for their favourite advertisement in general, *every single response* was based on a television ad. If young people are migrating to different platforms and activities, as we are so often told, then how on earth are these brands getting through to them?

LESSONS

The big lesson to be taken from all of the evidence laid out in this book is that we need to get out of this binary thinking – the zero sum game - which automatically assumes that for every success there is an equal and opposite failure and the rise of one platform or channel or service is automatically at the expense of another.

If we look at media from the ecosystem perspective, new entrants to a market may not just complement existing players but may also boost the whole market in ways that defy the doommongers. Technologies like the PVR are adopted not because of a desperate desire to avoid advertising so much as to watch more of the content we love. The growth in social media does not mean that word of mouth is the new television but rather it supports, reinforces, enhances and expands our viewing choices - much as it has always done, but more efficiently. The launch of mobile TV does not mean *"yet another competitor to the dusty old box in the corner of the room"* so much as freeing TV content to be viewed in new places and spaces. Increasing use of web TV to view on demand content does not compete with the linear schedules, but supplements the live TV audience and allows that audience to remain loyal to the programme. Growing penetration and usage of online technologies does not decrease viewing time but increases the total amount of time we spend with all media through media meshing.

This is a massive lesson for the media, marketing and digital industries to learn, because they have all been susceptible to the predictions raised above and have spent significant sums of money looking for the next big thing without realising it is often right under their noses. Companies like Pepsi have thrown out television in favour of the new kid on the block - social media – and witnessed a historically low market performance as a direct result. Thousands of companies have formed to steer us away from the 'tyranny of the linear schedule' with generally disastrous results. Even behemoths such as Google and Apple have struggled to gain a significant foothold in direct competition with the broadcast and platform brands and yet have gained huge amounts of traffic and revenues when sitting alongside and complementing people's core viewing experiences.

The main lesson for me, though, is that we ought to spend less time looking at what technology can do and more time looking at how and why consumers use it. We can then identify how technology can best be used to support and enhance that behaviour.

In this final chapter, I will draw on some of the lessons we have already learned (often painfully) to offer a few choice words of advice to each of the main industry sectors that are most likely to be affected by the opportunities arising from the third age of television. Whether or not they will heed them is a moot point. For many of them, the idea that TV will thrive in this future digital landscape is still difficult to grasp. Even for those who believe in television's place in 21st century marketing, there are still many bridges to cross.

Lessons for the broadcasters

Lesson 1 - Don't get careless or greedy

It's amazing that, with the online technology now available to us, some of the most lucrative innovations in television recently have come from some very old media ideas! For example, in its review of TV performance in 2010, Deloitte pointed out that one of the biggest-growing sources of on demand content was...DVD kiosks! Atoms don't always convert into bytes, as we have seen many times throughout this book.

Similarly, one of TV's fastest-growing revenue streams five years ago was not advertiser-funded programming, nor on demand IPTV services, nor even interactive revenues. It was people voting for outcomes in reality programmes, such as the winner of '*The X Factor*' or who should be kicked out of the '*Big Brother*' house. People regularly voted in their millions, primarily using a technology from the previous century; the telephone!

The revenues from premium rate phone lines rose significantly since 2000, bolstered by the increasing popularity of reality programmes and talent contests. The 2010 series of '*The X Factor*' in the UK alone generated more than 15 million votes, and a fresh revenue stream of tens of millions of pounds emerged. The beauty of these revenues was that they were wholly additional and required very little effort on the part of the broadcaster. What's not to like?

The problem was that the broadcasters got complacent. The phone voting revenues were seen as very much a revenue stream rather than a service that needed to be delivered. It was a potential recipe for disaster and resulted in one of the biggest crises UK television has faced for many years.

Between 2004 and 2006, a raft of voting scandals emerged, starting with *'The Richard and Judy Show'* and, in rapid succession *'X Factor'*, *'Strictly Come Dancing'*, *'Big Brother'*, *'Dancing on Ice'*, *'The British Comedy Awards'* (which subsequently went off air) and, perhaps most damaging of all, children's show *'Blue Peter'*. In the latter case, the show's producers decided they didn't like the name of the *'Blue Peter'* cat, voted for by viewers. Cookie, apparently, didn't fit the bill, so they changed the outcome of the vote so the cat would be named Socks.

Now, the naming of a cat is hardly an earth-shattering event, but the potential loss of faith in two of the most high profile and respected brands - The BBC and *'Blue Peter'* - is immeasurable. All for just a few million pounds.

The moral of the story is not to grab at an emerging revenue stream without understanding what it entails. There is no such thing as a free lunch. As we have already seen, trust is one of the most valuable assets TV companies own, but the breadth and depth of the phone voting scandal demonstrated that it could be so easily broken, with barely a thought to the consequences. Fortunately, judging by the latest voting figures, trust appears to have been restored...but only just!

As we move towards television's connected future, many more such temptations will come the broadcasters' way. Relaxed regulation – especially in the area between editorial and advertising – means that they will be tempted to include more commerciality into the programmes, either directly or subliminally. The ease of generating response will mean that transactions may be offered that do not necessarily benefit the viewer. The list of potential pratfalls and booby traps is very long. TV broadcasters need to become more adept at avoiding them.

Lesson 2 – Don't assume the status quo

The casual reader could be forgiven for thinking all was well in the world of television because relatively little has changed. Viewing, engagement and interaction have all increased incrementally. A theme running throughout this book is that change is evolutionary, not revolutionary, and that the major players of yesterday are still very much in place today.

Sometimes, the status quo hides change. Overall, after the audience declines of the 1990s and early 2000s, the four main terrestrial broadcasters in the UK have maintained a very steady share of the viewing audience, but that is all to do with the success of their digital channels.. We can expect to see that shift continue, which may pose a revenue challenge, as advertising rates for the digital channels are generally much lower than those for the flagship channel.

Channel brands are currently how people intuitively search for programmes to watch on demand and therefore the primary destination for the vast majority of the on demand audience. All of that <u>could</u> change if an aggregator could make it easier for people to find and access programmes. There are many waiting in the wings! Channel brands have been remarkably resilient so far (for example, as on demand destinations) but they need to continue to fight hard to maintain that position.

Similarly, the current strength of the linear schedule is quite remarkable, when you think of the enormous range of alternatives being pushed at viewers nowadays. Although I strongly believe that it will be at least a decade, probably longer, before total non-linear viewing hits the 20% share mark, I also believe that mismanagement of the schedules – and particularly the break lengths in between the programmes – could change things considerably. If, for example, broadcasters attempted to maximise short-term revenues by loading ad breaks with unwatchable direct response advertising, the audience would quickly find ad-free alternatives. Advertising breaks need to be 'curated' just as much as the programme schedules.

Lesson 3 – Orchestrate the media experience

The traditionally passive and one way nature of television viewing has meant that, until now, broadcasters only needed to worry about getting the best content on screen at the right times. Nowadays, the timing is becoming less important because of on demand, but the need to think beyond the screen has become paramount. Broadcasters need to think of the programming as the start point, not the end. They need to understand what audiences <u>may</u> want to do with the content beyond simply sitting back and watching it.

Orchestrating the viewing experience before, during and after the programme offers so many advantages, that it is no wonder the broadcasters are already active in this arena. Real-time apps linked to programmes like reality, soaps and sport can increase engagement with the programme itself and therefore loyalty to the series. They will increase the audience to the live broadcast, because that is how the media experience can be best enjoyed. It can promote the programme to people who maybe would not normally view, especially via social media. It can also offer advertisers new ways to integrate their brands and products into the TV content people love.

It will not be enough to simply attach additional content or interactivity to the programme itself. The orchestrated media experience needs to be simultaneous, relevant, intuitive and engaging. It needs to flow from audience research, but way beyond the usual ratings analysis or focus group approaches that have typified TV's traditional methods of understanding their viewers. In some ways, TV can learn from the online world by optimising customer experience or usability – not objectives that have been important to broadcasting over the years.

Sometimes, it can be enough to simply observe the audience in action. The example I gave from *'The Royle Family'*, highlighting the way they play a betting game around *'The Antiques Roadshow'* reflects how people like to play with TV (my own personal version is the betting syndicate my fellow 10 year old friends and I built around cartoon series *'Wacky Races'* in the 1970s!). Similarly, the 'Guess the Ad' games that people play in the commercial breaks, as we saw in the Thinkbox Engagement Study, offers some real orchestrated media opportunities to increase engagement with the advertising. The key is to make sure it is constructed from the consumer's point of view.

I'm aware that most of the examples I have used appeal to the playfulness of the television audience. Having observed how people really behave when they are watching TV in their own homes, this is one of the great unexplored opportunities. We've seen plenty of examples of how this appeal to the audience's willingness to play and compete can reap benefits. The BBC screened *'Test The Nation'* in 2002 and got 8 million people involved. Programmes such as *'Million Pound Drop'* have taken such play-along participation a step further. Channel 4 has appealed to this sense of playfulness in some of their themed ad breaks recently and advertisers are increasingly looking for ways to use gamification techniques to move TV viewers closer to their brands. But, whether it's through programmes or commercial breaks (although I suspect the most fertile ground will be the space between the two), we have only just touched the surface in terms of appealing to the audience's desire to 'play'.

It should be the broadcasters who lead the way here, as the thing most viewers want to play along with is their content. No doubt, the platforms will have a huge influence in enabling something of the scale to get millions of people playing along together.

Making best use of these opportunities will mean more integration between departments and with outside agencies and specialists to help conduct the orchestra. Many of the best examples of orchestration I have seen – such as the integration of games stats, betting and social media into sports programming – have added many layers of value to a traditional TV experience.

If the orchestra plays in harmony, it will almost certainly lead to greater audience involvement and more revenue opportunities.

Lesson 4 – maintain the quality lead

One of the main reasons that television has maintained its predominant entertainment position is down to the quality of the product - both delivery (in terms of big screen and high definition picture quality) and content (high production values, content created specifically for a lean-back, TV viewing context and a strong storytelling power). These factors have helped TV to maintain the live broadcast audience levels as well as expand its reach and viewing via the new platforms, through timeshift and placeshift viewing.

In terms of delivery, broadcasters could look beyond HD to Ultra Definition. This not only improves the quality of viewing but also creates even more of a strain for IP networks, meaning the more efficient economics of broadcast should continue. Although the technical experts assure us that, in future, the IP infrastructure should have no problems coping with all TV being delivered via IP, the difficulties the current infrastructure has with the relatively light existing traffic suggests that day is some way off.

Rather, it is through quality of content that TV can continue to maintain a clear lead over other forms of entertainment. The more efficient production capabilities of digital technology means all kinds of genres can now capitalise on production values that would have been impossible to replicate just a decade ago, and it is vital the TV industry exploits them to the full to create a clear gap between what the online community can create via UGC (user-generated content) and SPUG (semi-professional user-generated content). People still have an expectation that, if it has been on TV, that is a guarantee of quality.

Lessons for the Content Producers

Lesson 1 - Get closer to advertisers

As we have already seen, the history of close collaboration between programme content producers and advertisers has been somewhat patchy. Certainly, the experience of advertiser funded programming will have deterred many production companies, but they may have to get used to it, as programme budgets come under strain and advertiser money increasingly seeks to get closer and closer to the content the viewers are there to see - the programmes.

There is a long history of marketing money supporting editorial media, in print, broadcast and now online. Television has been most prominent in separating the two, mainly due to tighter regulation. But as regulation loosens and the definitions of where a programme begins and ends become more blurred, then we can expect to see more advertiser money flowing into the programme production process.

It will be around those blurring lines that these collaborations could be most effective. As with AFP, it will mean working across and between the advertiser, the agency(s) and the broadcaster, so it is never going to be easy. Following the money rarely is.

Lesson 2 - Slicing and dicing, bending and extending

The T-mobile 'Life's for Sharing' campaign is a fantastic example of the idea leading not just the media, but also the creative.

T-mobile and their agencies created a great piece of audio-visual content, which they then tried slicing and dicing, bending and extending so that it was the driving force in the production of over one hundred different creative executions. Examples include;

Slicing the content into bite-sized pieces, often just a few seconds long, which were seeded into viral networks to spark interest and create easy sharing of the experience.

Dicing the content into new TV ads, integrated with a direct response message, to elicit the positive sentiments towards the original ad in order to persuade people to engage and respond.

Bending the content into new directions; for example, for a PR event, T-mobile recruited the judges from '*Strictly Come Dancing*' to judge the performances of employees performing the Life is for Sharing dance moves.

Extending the content to fit the context - most notably with the launch ad, where T-mobile took over the whole commercial break with the full-length two and a half minute ad featuring the whole dance.

All TV content needs to be approached in this way. TV producers need to think like T-mobile. Even though the brief may be for a drama series or an adverting campaign, the first thought should be 'what content will we have?' It might be extra footage, spoof edits, promotional clips or live commentary. It may be for promotional purposes, second screening, additional programming or exclusive access. The future of television will always be about millions of people consuming the same core content, most of them simultaneously. Each will create their own personal experience out of that content, based on the various ways it is sliced, diced, bent and extended. The more opportunities they are offered, the more of them there will be, and the more engaged they will be as a result.

Lesson 3 – Synchronise content!

There has long been a complaint from the production community that broadcaster revenues are increasingly failing to keep pace with the actual costs of production, often resulting in production companies getting more rights to their content beyond broadcast. Exploiting the second screen and creating orchestrated media experiences around that content is one of the major opportunities.

As second screening grows in both popularity and functionality, the ways in which the viewing experience can be enhanced and extended are growing apace. Live stats to support sports viewing; information about the cast of the prime time drama we're watching; play-along competitions or chat with others within that programme's community of interest. These are all becoming reality, adding value to the viewing experience. As more and more advertisers enter this space (and why wouldn't they?) then new revenue opportunities could make this a lucrative path for independent producers to follow.

Lessons for Marketers

Lesson 1 – Fame + emotion + a good story

When the IPA unveiled the results of their meta-analysis, '*Marketing in the Era of Accountability*' the authors described the common wisdom that it was rational messages that shifted products as *"fundamentally wrong"*. Campaigns based on fame and emotion that were most successful at shifting product.

Of course, in reality, it is difficult to separate the two. Campaigns that engage us emotionally will naturally be the ones we want to talk about and share. The recent John Lewis Christmas campaigns pulled every heartstring imaginable while also generating millions of YouTube hits and bucket loads of conversation. Is it a fame campaign or an emotional one? Either and both - it is a fantastic example of the two working together.

Of course, both work best with a really good story. We have seen how storytelling is ingrained in us from birth and the most effective way of moulding memory and changing behaviour. The emerging ability to mould and shape the brand story across many different media experiences adds a new dimension to modern day marketing. Transmedia storytelling will almost become generic through the wide array of different ways to experience the story.

The big idea will set off a barrage of touchpoints aimed at bringing consumers closer to brands. But any big idea needs a narrative spine that can only be delivered with the full audio visual experience and across a large enough population to gain social currency. Given TV's remarkable resilience, it would be a shame if advertisers failed to take advantage of TV's ability to provide that narrative spine to their future transmedia storytelling.

Lesson 2 – We need to talk

TV is driving conversation more than ever before, enhanced by the power of social media.

The big advertisers who are laying claim to the social media space are also spending more on TV. Increasingly, it appears that more advertisers understand that TV and social media drive each other.

Most important of all, though, don't forget that most of those conversations will continue to happen below the radar, face-to-face or by telephone. That is where most of our social lives take place, and TV influences them more than we ever realised.

Lesson 3 – TV is part of a thriving ecosystem

Most media agencies these days believe in integrated media

planning, even when most of the trading is conducted in silos. They recognise that consumers use many media, often simultaneously, and experience brands and content via a bewildering array of channels. We can't simply separate 'online' and 'offline' media any longer. We are seeing the emergence of a new media ecosystem and TV is very much at its centre.

Today's consumers live two lives; in home and out of home. Each offers a very different context and experience.

Our out of home time encompasses commuting, work, shopping, socialising, sightseeing and simply getting on with our busy lives. Media – and brands – are everywhere, clamouring for attention, along with work deadlines, documents to study and queues to navigate. A great deal gets through – a perfect illustration is through the way that UK celebrity "mind reader" Derren Brown taps into the massive amounts of parallel processing in our brains to steer our subsequent behaviour[71].

Television has a big part to play in our out-of-home media life. It might be live football in a sports bar, watching last night's drama on the way to work or a sneaky peek at last night's highlights from *'The Apprentice'* in our lunch hour. It could be a conversation about Channel 4's new comedy on Facebook the next day or catching the news headlines on a Sky News screen at the station. It doesn't matter. Within the out-of-home ecosystem, TV is one of many channels competing for our attention. It competes with lots of other activities. Of course, it is progress for TV to even be considered to be part of the out of home media landscape; it was almost completely absent from it just two decades ago.

It is the role that television continues to take when those same consumers are in the place they (usually) love the best that is where the bulk of its power, value and resilience lies. It is the place where we can stop being attentive and start getting engaged. It is the place where entertainment is most demanded and advertising is at its most effective. It is the place called home.

This is the place where consumers congregate in a room that is designed around that electronic fireplace we call the television set. They are generally more relaxed and happy. They have established routines often built around the TV schedules and their own family rituals. It is often referred to as quality time. But, despite its dominance, TV is still part of an eco-system, especially because of the high adoption of smart, portable and connected TV and companion screens. This is the ecosystem that the 'Tellyporting' research described so well and which now brings television much closer to the point of purchase.

As we have seen throughout this book, that creates some exciting opportunities.

Lesson 4 – It's big ideas, not just big pipes

There has been a trend recently for marketers to claim to be putting aside significant budgets to test new advertising ideas. Most of these have been based on the technology (online) and/ or the channel (e.g. social media). The low costs of entry and availability of data means it is a great way to try out new ways of doing things with low risks attached. Some great campaigns have come from such experimentation and it should be applauded. So why am I so underwhelmed by the approach taken so far?

It's the point about transmedia storytelling again, where there has been more focus on the transmedia than the storytelling, and yet it is the storytelling that really delivers the goods. There has been so much evidence recently about the power of good creativity to shift product, in far more significant numbers than any of these new channels have been able to demonstrate, that it seems a shame that similar budgets aren't available to experiment with the big ideas that go to the heart of great marketing, regardless of the technology.

If some of that 'digital' money is really going to allow for such experimentation to take place in order for those big ideas to really show what they can do, then it will be money well spent. I also believe it will result in money coming back to TV, because the best ideas need the impetus TV can bring to maximise the returns.

Understand what the technology can do but don't see it as an end in itself. It's the big ideas and engaging creativity that shift product, not the pipes they flow through.

And finally...

Lessons for the Digital Industry

Lesson 1 – Be evangelical, not fundamentalist

At Thinkbox, we often made the distinction between 'evangelists' and 'fundamentalists'. We were evangelists, we felt, in that we loved TV and wanted to share our understanding of how it worked, but not to the detriment of anything else. An evangelist can also see the good in other ways of doing things and nowhere was this more apparent than in our equal love for all things online. It is only by being an evangelist that you can see the wider picture, and understand that it is not about competition but complementarity.

Fundamentalists, on the other hand, can only see the one way. Everything else is inferior, impure or just plain imperfect. For the fundamentalist, it is their way alone that will prevail, overwhelming all alternatives through its basic, pure goodness.

I would say pretty much all of the '*TV is dead*' predictions came from digital fundamentalists, who simply couldn't see any other possible future.

The difference between and evangelist and a fundamentalist is, I believe, humility; the understanding that there is more than one path to follow with room for us all. Sometimes that humility is a result of one's own bout of soul searching; certainly TV lost much of its traditional arrogance as a result of the crisis of confidence in the mid-2000s. Sometimes, it is a necessity for survival.

If there is one segment of the communications industry I hope would learn from this book, it is the digital fundamentalists (who, by definition, are the least likely to read it). Learn from the reasons <u>why</u> TV has defied your predictions of gloom and has thrived as part of the digital landscape rather than been overwhelmed by digital competition. Those reasons are a great deal less to do with technology than they are to do with human needs states, evolution and established behaviour patterns. These are not things to fight against, they are part of the ecosystem, an environment that new technology has generally sustained rather than destroyed.

Learn from that experience, and the technologies the online industry promotes can become even more integrated into our living rooms and into our lives outside it.

Lesson 2 - Stop thinking binary – how to enhance rather than conquer

Digital is perfect - perfectly defined, perfectly reproduced, perfectly maintained and it can be managed to become perfectly integrated and perfectly seamless. It is exactly what it says on the tin, precisely the sum of its parts. It is clear, rational and totally transferable. The ways in which the pioneers of the industry have understood that, whether via the ubiquity of Microsoft, the infrastructure of Apple or the efficiency of Google is really what sets them apart. When that perfection is realised (and there are numerous examples where it is not), it is a beautiful thing, and will affect our future more than we can ever know.

But digital perfection is driven by a binary technology that has also spawned a great deal of binary thinking; as we have seen recently with classical economics, such thinking can fail to keep pace with the complexities of real life.

246

Binary thinking is about winners and losers, in and out, predators or victims. It is about believing one thing can only thrive at the expense of another. Digital binary thinking has echoes of classical economics, with the naive belief that the rational, functional benefits of technological innovation will be immediately perceived, understood and acted upon. I have lost count of the number of business start-ups I have seen that have struggled because of that belief. Markets don't act in binary ways, and neither do human beings.

An ecosystem approach would have perceived a whole greater than the sum of the parts, such as total TV viewing rising with each successive online innovation. It would have been much quicker to exploit the amazing advantages from providing complementary (rather than competitive) content to the television set. It might have seen online brands like Google and Apple become more integrated into the TV viewing experience, one of the few areas neither company has so far been able to dominate. The lessons in this book help us to understand why that is still the case.

Lesson 3 – understand the TV mindset.

As we have already seen, the mindset when people sit down to watch TV is generally "entertain me!", and TV broadcasters have been rising to that challenge for more than 60 years. It is a mindset that has been shaped by the stresses and demands of contemporary life; TV audiences don't want to engage their brains or make decisions any more than they need to. If we are all cognitive misers most of the time, we are veritable Scrooges when it comes to applying mental effort once the working day is over.

This has been a major challenge for an online industry that has thrived on a totally different mindset; a mindset that responds to the simple question '*what do you want?*'. This has been the mindset that has fuelled search, shopping, browsing and entertainment on the web since dial-up modems. It works extremely well when people are on a clear, goal-directed journey. It works less well when people are trying to find entertainment without a clear idea of what they want OR an already established viewing routine. '*What do you want?*' is far too challenging when we are in this mindset.

There has been talk recently of Facebook attempting to become our social electronic programme guide. The argument goes that a combination of friends' recommendations and algorithms will provide all of the data needed to provide people with the perfect range of options for what to watch, enabling them to manageably find the programmes they will learn to love (if they don't already) via what their friends love and what the data tells us they will find attractive. This approach would certainly help to provide a degree of manageability over the sheer amount of choice now available, and it may well provide an alternative (or complement) to the existing electronic programme guides BUT it would also need to account for issues which are not currently part of online search or recommendation, such as shared viewing, viewing flow, existing appointments to view and the desire for serendipity or discovery; the 'surprise me' factor that all good television schedules should contain.

These are not concepts which the online world has so far managed to fully grasp, as they tend to go against the principles of personalisation, open access, equality of content and choice maximisation. However, they are fundamental to how people have made their viewing choices since broadcasting began and are based on clear needs states which won't be changing any time soon. Why should they? They have been serving the viewer well throughout some of the most tumultuous times in broadcasting history.

Lesson 4 – embrace TV – it's your (potential) best friend

It has been difficult for the online world to accept that television is here to stay; as we have seen throughout this book, the predictions of TV's demise has often been based mainly on wishful thinking and an outdated replacement theory approach. It is difficult to turn around and make best friends with a self-made enemy.

There is room for multiple winners in the ecosystem approach. Between them, TV and online take up more than half of our waking hours; they do very different things in highly complementary ways. If they continue to work together, not cannibalising but enhancing, then both activities can become more relevant, entertaining, connected and ultimately profitable. We are already beginning to see the results.

Lesson 5 – the world may be digital...but people are analogue

I began this book looking backwards; through my own personal and career history, through television's history and, finally, through the story of television's presumed demise. The death of television was not to be a silent, dignified affair, but accompanied by a cacophony of noise about the virulence of the symptoms and the imminence of the patient's departure from this new, efficient, democratic, *accountable* digital world.

Television was supposed to have been all but dead by now because many wrong assumptions were made.

It was assumed that consumers were just as interested in the gadgetry and functionality of the digital revolution as the digerati whereas, in the living room at least, they just want to be entertained as easily as possible.

Everything we have learned about how people think, feel, decide and behave has been turned on its head in the past decade or two and, as a psychologist, I have been excited by the results. People aren't mechanical, logical, predictable or rational. They are complex, organic, emotional and seemingly random. People are essentially analogue in nature, but the very things that make them greater than the sum of their parts, have been lost in the analytics.

This is an issue I am planning to tackle in my next book, but in the meantime let's just remember that consumers never have behaved in the rational, self-interested, predictable ways that both classical economists and digital experts have had us believe; it is just that now we have the tools to measure, understand and optimise the ways that consumers...sorry, human beings...really do interact with technology and with each other in this brave new digital age.

Final Thoughts

Over the course of my career, I have had an unbelievable opportunity to provide the research and insights that help us understand the factors behind television's historical dominance, its current resilience and its future opportunities. In doing so, I have tried to emphasise the issues and influences that will shape television's digital future and map out its place in the emerging media landscape. Overall, because this perspective is based on fundamental consumer needs rather than more changeable technological developments, I have concluded that television is in a good place, ready to profit from the changes that are coming and maintain its predominant role in the media ecosystem. The future is bright.

The future is never guaranteed, though, and there are still some major challenges that television will need to negotiate in order not to throw it away (and it is a very thin line between success and failure in this emerging landscape).

The broadcasters do face an uncertain advertising market and there is no shortage of 'video views' on offer to keep downward pressure on their rates. It is a thin financial line between a schedule strong enough to maintain its current live audience and find new ones through timeshift and on demand to a schedule that feels a bit tired and vulnerable to thousands of alternatives, many just a few twitches of the thumb from the screen. How advertising fits into this environment will also need to be carefully managed. Even if the content is 'must have' it can be found through other means, advertising-free.

Marketers also need to take heed. I think in the rush to kill off television, we forgot just how strong a hold it continues to have on our lives. The viewing experience, especially during peak time, is a powerful phenomenon and fertile ground for brands to work their magic. The more creatively they weave their spells, the more their customers will love them; and that is what sells. Why would you wish it away?

The platforms will need to keep their subscribers within the walled garden as much as they possibly can. They don't need to offer more choice; they need to offer better ways for their subscribers to make best use of the choice already available to them. Over the next few years, there will be more and more services fighting for a share of their time, and most of them will offer for free what has traditionally been part of the paid-for package. As it reaches maturity, the pay TV industry needs to concentrate on keeping those customers satisfied enough to continue to hold on to the cord.

Content providers must continue to think beyond traditional television formats in order to make great television. Producers of both programmes and commercials (and everything in between) must offer viewers as many ways as possible not only to view, but to experience. Whatever the big idea or high concept, it must make great television. The audience is ready to be engaged, but is equally prepared to switch. So far, they have shown a real enthusiasm for innovation and creativity in television. For now, the audience is there in record numbers and amazingly accepting of your presence in their lives. Only you can change that.

For government and regulators, it is difficult to know where to start. Television has always been one of those things that the UK does well, partly because we have the good-fortune to be English-speaking and partly because one of the best examples of 'big government' – the BBC – has helped create high programme standards and a commitment to innovation; I don't believe BSkyB would have been half as innovative as they have been in the last 15 years if the free alternative, as defined by BBC Television, hadn't been so compelling. Don't unnecessarily limit or destroy what is a fragile industry, one that is still responsible for billions of pounds worth of trading exports and an enhanced cultural standing in the world. But don't treat it as a special case, either. It can be used positively, to bridge the digital divide and engage the national psyche.

I won't restrict my final comments to the digital industry, because we are all digital now. Instead, I'll reflect sadly that, despite the massive weight of evidence to the contrary, television is, according to some at least, still dead. Even as I write, new examples of that binary thinking emerge. I have many to choose from, but I'll select this quote as it was uttered by the former chairman of Channel 4, Luke Johnson, at the 2011 IBC Conference in Amsterdam;

"It's the end of television as we know it. The television in the corner of the room is medium-term history. Traditional broadcasters are in the business of managing decline."

Meanwhile, the US digital industry proclaimed the death of the network schedules, based on 4 weeks' worth of incomparable data.

We have been hearing stories about the death of television now for almost 20 years. We have been promised a world free from the shackles of the programme schedules, free from channels themselves, free from the need for payment of any kind and free from all of that nasty advertising. The predictions have been relentless and compelling. They just haven't been true. And the reason they haven't been true is that we were looking in all of the wrong places.

We were getting hot under the collar about millions of views on YouTube and completely discounting billions of views on commercial TV. We were sweating over search, the new god of effectiveness, and failing to see how much television was driving it. We assumed social media would compete for TV time, not complement it. We assumed so much because we became bedazzled by the analytics and ignored the insight.

In this on demand, personalised, algorithmic, vacuum-packed future that the consultants and technology suppliers have been predicting, there was no thought given to how real consumers think and feel, and what drives their behaviour. They want to share (in the real world, if possible). They want to be able to trust in the face of an increasingly 'anything goes' world. They want to experience great stories, curated and presented by trusted editors. They want to play, but without too much effort involved. They want a social currency they can share and shape. They want to be surprised and immersed and sometimes mesmerised. Television has always provided that.

Now people also want TV to be bigger and better. They want to take it with them wherever they are and through whatever devices they have available. They want to supplement their viewing and, sometimes, to interact with the content, often respond or even purchase. Television in its third age is increasingly offering all of that too. Television can be whatever its viewers want it to be...but most of them want it to be television.

References

(Endnotes)

1   Nicholas Negroponte – *'Being Digital'*, Alfred A Kopf Publishing 1995

2   Data from Boston Consulting Group's 'The Connected Kingdom' report, sponsored by Google – October 2010

3   Statistics from OFCOM, RTL Key Facts 2011, Google's *'Digital Britain'* report

4   Daily Telegraph 21st September 2009. http://www.telegraph.co.uk/finance/newsbysector/mediatechnologyandtelecoms/media/6622875/Final-farewell-to-worst-deal-in-history-AOL-Time-Warner.html

5   Source – OFCOM Media Report 2011

6   George Gilder – *'Life After Television'* – W. W. Norton & Company, 1992

7   Chris Anderson – *'The Long Tail'* – Random House 2007

8   Source: NARM/NPD Research, 2011

9   'The Hollywood Economist – The Hidden Financial Reality Behind the Movies' by Edward Jay Epstein, published by Melville House 2010

10   Source – Eurodata 2011 (Mediametrie) as published in RTL International Key Facts 2012

11   Source – Thinkbox 'Tellyporting' study reports catch up accounts for 89% of the reasons for on demand viewing, up from 78% from the 'Me-TV' study three years earlier

12   Source – BARB 2010

13   The recent media behaviour research monitor, covering all media use and launched by the Institute of Practitioners in Advertising (IPA) in 2007

14   Elizabeth F. Churchill – *'Enticing Engagement'* – Interactions May/June 2010 for a longer list of online behavioural measures of engagement

15   Thinkbox 'Brainwaves' Study (http://www.thinkbox.tv/server/show/nav.1367), Mindshare '3 Screens' Study (http://www.thinkbox.tv/server/show/nav.1292),

16   As (iv)

17   Robert Heath – *'The Hidden Power of Advertising: How Low Involvement Processing Influences the Way We Choose Brands'';* ADMAP Monograph, March 1999

18   Write-up of the study can be found on the Thinkbox website - http://www.thinkbox.tv/server/show/nav.854

19   Malcolm Gladwell – 'Blink', Penguin Books 2005

20   Antonio DiMasio – *'Descartes Error – Emotion, Reason and the Human Brain',* Random House 1994

21   Unpublished, but further details can be obtained from Dr. Alistair Goode at Duckfoot Research

22   Rory Sutherland interview – *'Questioning the Nature of Research'* – Research Magazine August 2011

23   Les Binet & Peter Field – *'Marketing in the Era of Accountability'.* IPA 2007

24   Professor Richard Silberstein – *'Long-term Memory Encoding and Brand Preference'* International Journal of Advertising, Vol 27 no. 3

25   As above

26   Geoffrey Beattie, Professor of Psychology, University of Manchester – *'TV & The Brain'* (from ITV)

27   William Goldman – *'Adventures in the Screen Trade'*, McDonald & Co. Ltd. 1984

28   David Brennan – *'Influence Through Storytelling'*, ADMAP October 2011

29   Campaign September 2008. "*Mildenhall Hails Power of Storytelling*"

30   Including studies by Carat and Granada TV in the 1980s, Turner Broadcasting in 2009,

31   As reported in Bloomberg Business Week – 'This is your brain on advertising', 8th October 2007

32   As reported on Thinkbox website http://www.thinkbox. tv/server/show/ConCaseStudy.1141

33   Thinkbox *'TV Together'* Study, 2010, BARB analysis conducted by RSMB

34   IP Deutschland -

35   Television Opinion Monitor, 2007, IPSOS

36   Ed Keller & Brad Fay – *'The Face to Face Book – Why Real Relationships Rule in a Digital Marketplace'* published by Free Press, 201

37   A full video of the speech can be found at http://www. youtube.com/watch?v=Ry1ropGpx-k

38   Mark Earls – 'Herd*: How To Change Mass Behaviour By Harnessing Our True Nature'*, Wiley Publishing 2007

39   Ray Snoddy – Mediatel Newsline article, 14th September, 2011

40   A full case study can be found in the 2010 edition of *'Advertising Works'*, the IPA's compendium of successful Advertising Effectiveness Awards winners

41   'Sponsorship: A Brand's Best Friend' – Thinkbox 2008 (http://www.thinkbox.tv/server/show/nav.1036)

42   Neil Strauss – *'The Game: Penetrating the Secret Society*

of Pick-up Artists' Regan Publishing 2005

43   E.g. Robin Wight – '*The Peacock's Tail and the Reputation Reflex: The Neuroscience of Arts Sponsorship*' published by Engine 2007

44

45   Price Waterhouse Coopers – '*Advertising Payback – Is TV Still Effective*' – May 2008

46   Thinkbox 'Payback 3' study – on Thinkbox website '*Payback 3: Ad Success in Tough Times* (http://www.thinkbox.tv/ server/show/nav.1818*)*

47   Paul Dyson – '*Cutting Adspend in a Recession Delays Recovery* – WARC Online, May 2008

48   Mediacom analysis of database of response campaigns – detailed in The New Rules of Response' – Thinkbox/Mediacom – on Thinkbox website (http://www.thinkbox.tv/server/show/ nav.1245

49   'TV & Online – Better Together' – Thinkbox/IAB – on Thinkbox website (http://www.thinkbox.tv/server/show/nav.1019)

50   The Best Organisation – '*Online Journeys*' – January 2010

51   Nielsen Ad Dynamics – analysis conducted by Thinkbox, July 2009

52   'The New Rules of Response' – Thinkbox/Mediacom – on Thinkbox website (http://www.thinkbox.tv/server/show/ nav.1245)

53   Report on BBC News website by Anthony Reuben – '*The Recession's Unexpected Winners*' – 12th February, 2009

54   Amanda D. Lotz – '*The Revolution Will Be Televised*', New York University Press, 2007

55   Based on BSkyB Company Report year to June 2012

56   Source GFK/Deloitte 'Why TV?' research for Edinburgh International TV Festival 2012 – 6% of UK consumers have a TV

set with 3D capability

57   Jack Myers – *'The Myers Report'* – April 2007

58   J. Clift et al – 'The Exploratory Study of High Definition Advertising and Consumer Response', ADMAP February 2007

59   For a recent demonstration, go to http://en.wikipedia.org/wiki/Technologies_in_Minority_Report

60   British Video Association – BVA Yearbook 2011

61   Screen Digest 2008 (when boxed sets achieved 13% of total DVD market) plus later reports on continued growth to extrapolate to 15%.

62   "Any references to Sky and its service in either pen portrait are fictitious"

63   David Brennan – *'TV Programme Sponsorship – A Brand's Best Friend'* – published in Journal of Sponsorship, Vol 2 No. 2, February 2009

64   Peter Pynta – 'Brands Compete for Emotional Survival' on the Neuro-insight website (www.neuro-insight.com)

65   Peter Field – *'The link between creativity and effectiveness. The growing imperative to embrace creativity'* published by IPA, June 2011

66   Source BARB - *'Trends in Television'* available on BARB website (www.barb.co.uk)

67   Thinkbox - *'The Secret Life of Students'* – on Thinkbox website (http://www.thinkbox.tv/server/show/nav.852)

68   Based on BARB viewing data (2011) plus data from Touchpoints, Comscore and UKOM (2011)

69   Thinkbox – 'Generation Whatever' research- on Thinkbox website (http://www.thinkbox.tv/server/show/nav.851)

70   Deloitte UK – *'TV+ - Perspectives on TV in words and numbers'*, August 2011 – on Deloitte website (http://www.deloitte.com/view/en_GB/uk/industries/tmt/5d44f5fe4e0f1310VgnVCM2000001b56f00aRCRD.htm)

71   Two good examples are from Derren Brown's Channel

4 series *'Mind Control'* – *'Animal Heaven'* demonstrates how he influences two advertising creatives to produce an ad he had already guesses (http://www.youtube.com/watch?v=1UpUcgPP-YY), and *'Hamleys'* shows how he influences a choice of a single toy from almost a million on display at Hamley's toy store.

www.ingramcontent.com/pod-product-compliance
Lightning Source LLC
Chambersburg PA
CBHW030821090426
42737CB00009B/813